KT-177-842

Basic Biochemistry

Basic Biochemistry

A visual approach for college and university students

Dr. J. Edelman
Director of Research, RHM Limited
and previously Professor of Botany in
the University of London

Dr. J. M. Chapman
Lecturer in Biology, Queen Elizabeth College, London

HEINEMANN
EDUCATIONAL

Heinemann Educational Books Ltd,
Halley Court, Jordan Hill, Oxford OX2 8EJ

OXFORD LONDON EDINBURGH MADRID
ATHENS BOLOGNA PARIS MELBOURNE
SYDNEY AUCKLAND SINGAPORE TOKYO
IBADAN NAIROBI HARARE GABORONE
PORTSMOUTH NH (USA)

ISBN 0 435 60230 6

First published 1978
91 92 93 94 95 17 16 15 14 13 12 11 10 9 8

Printed in Great Britain by
Athenaeum Press Ltd, Newcastle upon Tyne.

Contents

Preface

Organisms function. In order to do this, they are organised. Anything which functions is organised, whether it is a motor car, a bacterium, the human body, a telephone network. Events which occur in them do not happen randomly but are directed according to a pattern.

It is a basic character of all organised systems that they require:

1. Structure – the structure carries the pattern which directs the function.
2. Energy – the energy is required to make the system function, or work, and also to build it and keep it in repair.

In organisms, the energy source – often carbohydrate or fat – is frequently used as part of the structure, and vice versa. So, sugars are a major energy source but are also used as the building material for plant cell walls; fats are also respired to produce energy but are part of membranes found in cells; amino acids are important units of the structural protein of skin and connective tissues but may also be used as an energy source.

These inter-relationships make the sorting out of the structure and function of living systems very complex. Their patterns are perhaps more easily grasped visually than descriptively. This book attempts the visual approach to biochemistry rather more than most others. Here and there it may labour the basic concepts more than many biochemists might think warranted. But the experience of the authors leads them to believe that many students, who are not making biochemistry their major study, often learn superficial complexities rather than basic simplicities.

For this reason, the book is aimed not only at biochemists, but also at those students who are about to embark on the host of subjects which demand some biochemical knowledge. These range from home economics to agriculture, from botany to medicine. It is not meant particularly for students at universities nor those at technical colleges or polytechnics, nor for very bright sixth-formers taking S level. We, the authors, however, hope that it will be found of some use as a groundwork by some students among all of these classes.

J.E.
J.M.C

Part 1 Compounds in the Cell

AMINO ACIDS AND PROTEINS

- Amino acids are joined together in chains to make proteins.

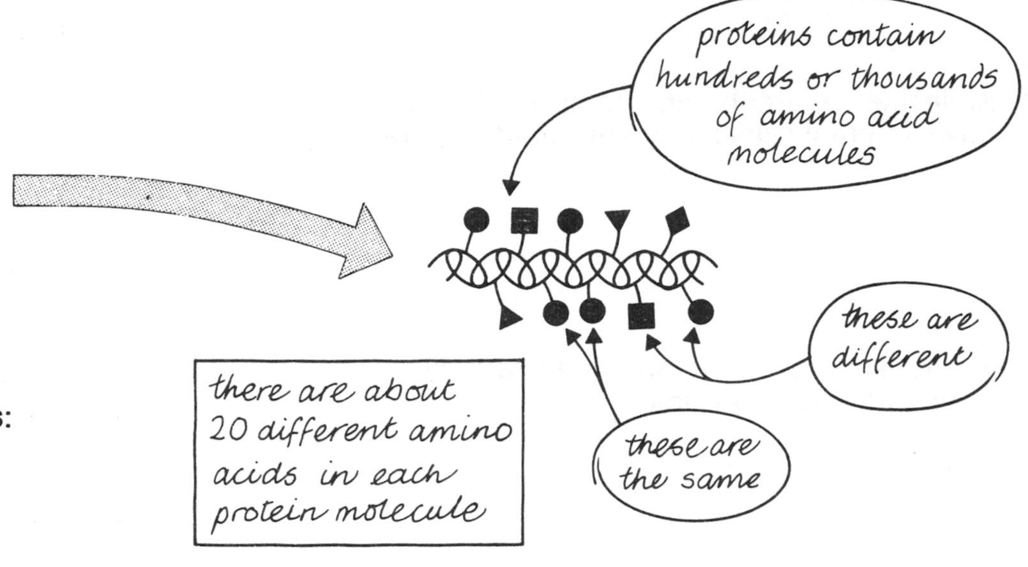

proteins contain hundreds or thousands of amino acid molecules
these are different
there are about 20 different amino acids in each protein molecule
these are the same

- The 20 amino acids which are found in proteins are called the **monomers**: all proteins are **polymers**.

- The amino acids are made of C,H,O and N (except 3 which also contain S).

for example

$C_2H_5O_2N$	glycine
$C_6H_{14}O_2N_2$	lysine
$C_3H_7O_2NS$	cysteine

- Each amino acid has an acidic group and a basic group in its molecule. These groups are attached to the same C atom.

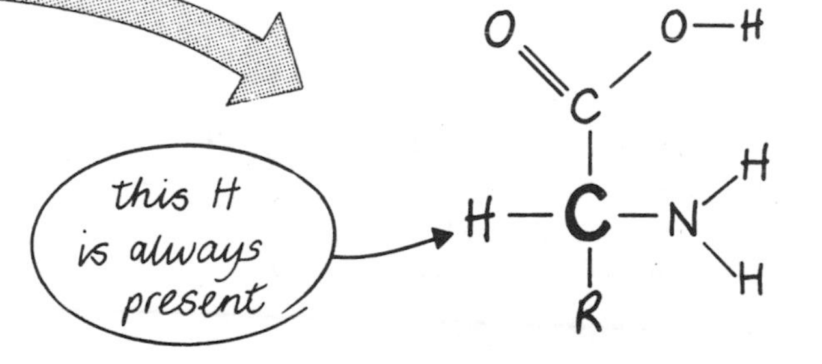

this H is always present
O
O—H
C
H—C—N
H
H
R

acid group (carboxyl)

basic group (amino)

- Compounds whose molecules have both basic and acidic properties are called **amphoteric** compounds.

- Molecules of amphoteric compounds may form 'internal salts'; they are called **zwitterions**.

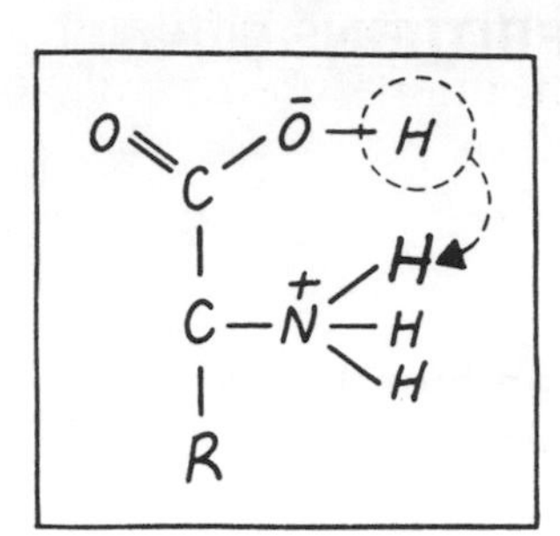

- Amino acids are different because they have different **R** groups.

H
R — C — COOH
NH_2

this is called the amino acid side chain

can be one of about 20 different structures

- Apart from glycine (in which the R group ≡H) all other amino acids have an **asymmetric** carbon atom. This means that they can exist in two different 3-D structures.

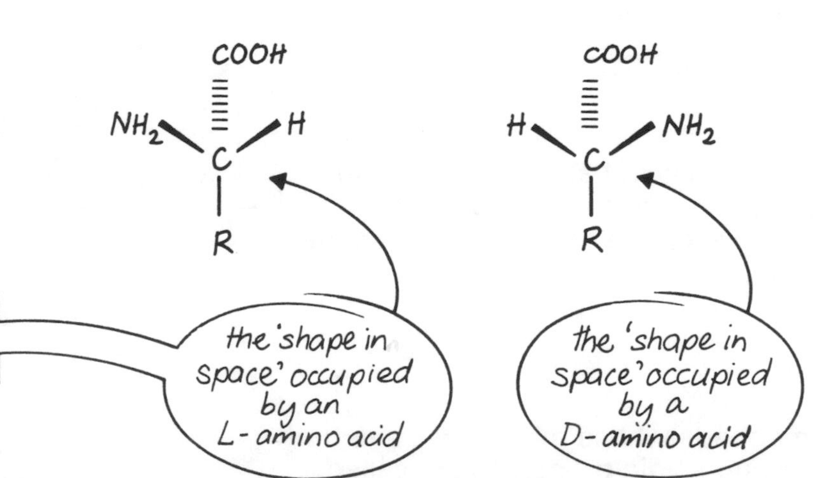

COOH
NH2
H
C
R
COOH
H
NH2
C
R
the 'shape in space' occupied by an L-amino acid
the 'shape in space' occupied by a D-amino acid

living organisms utilise almost entirely those amino-acids called L-amino acids

- We can put amino acids into 4 different classes depending on the nature of their side chains.

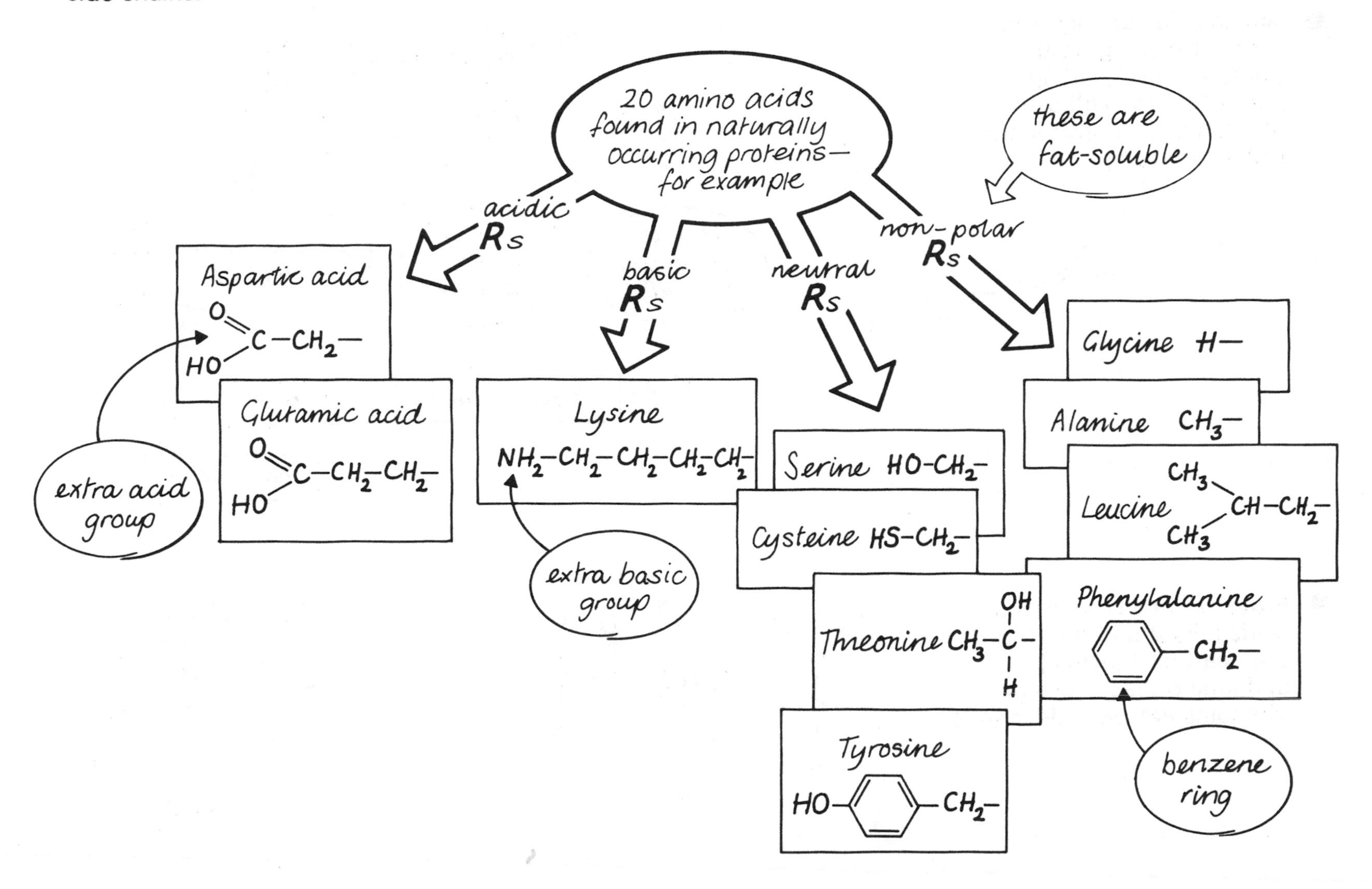

20 amino acids found in naturally occurring proteins—for example
these are fat-soluble
acidic Rs
basic Rs
neutral Rs
non-polar Rs
Aspartic acid
O
C—CH2—
HO
Glutamic acid
O
C—CH2—CH2—
HO
extra acid group
Lysine
NH2—CH2—CH2—CH2—CH2—
extra basic group
Serine HO-CH2—
Cysteine HS—CH2—
Threonine CH3—C—
OH
H
Tyrosine
HO—
—CH2—
Glycine H—
Alanine CH3—
Leucine
CH3
CH—CH2—
CH3
Phenylalanine
—CH2—
benzene ring

- Amino acids are not only used for making proteins. Some are also starting points for the synthesis of other compounds in the cell – for example:

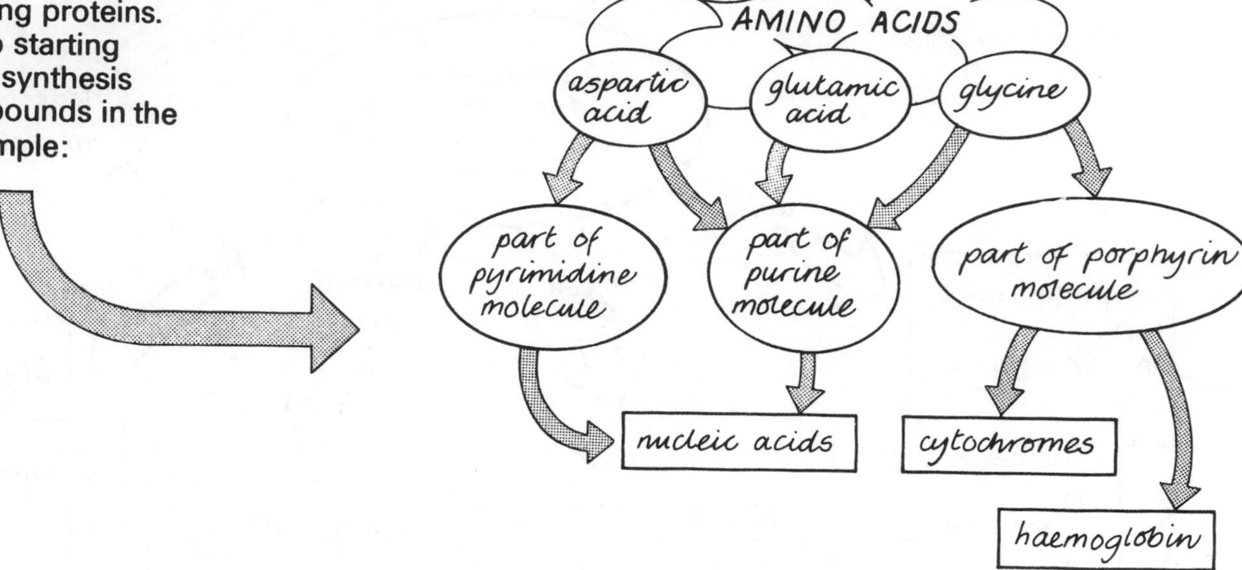

AMINO ACIDS
aspartic acid
glutamic acid
glycine
part of pyrimidine molecule
part of purine molecule
part of porphyrin molecule
nucleic acids
cytochromes
haemoglobin

- Proteins consist of amino acids joined together by their carboxyl (acidic) and amino (basic) groups. The joins are **peptide bonds**.

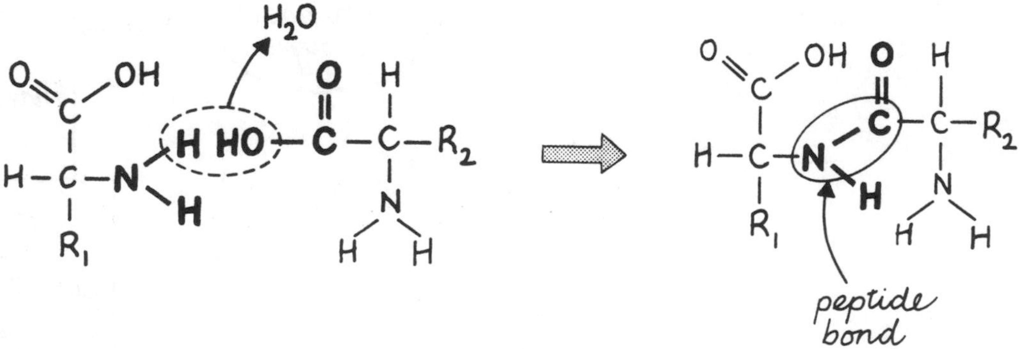

H_2O

peptide bond

- So the protein looks like this:
- There are lots of different proteins. Different proteins contain different combinations of amino acids like letters in (very long) words.
- Some examples:

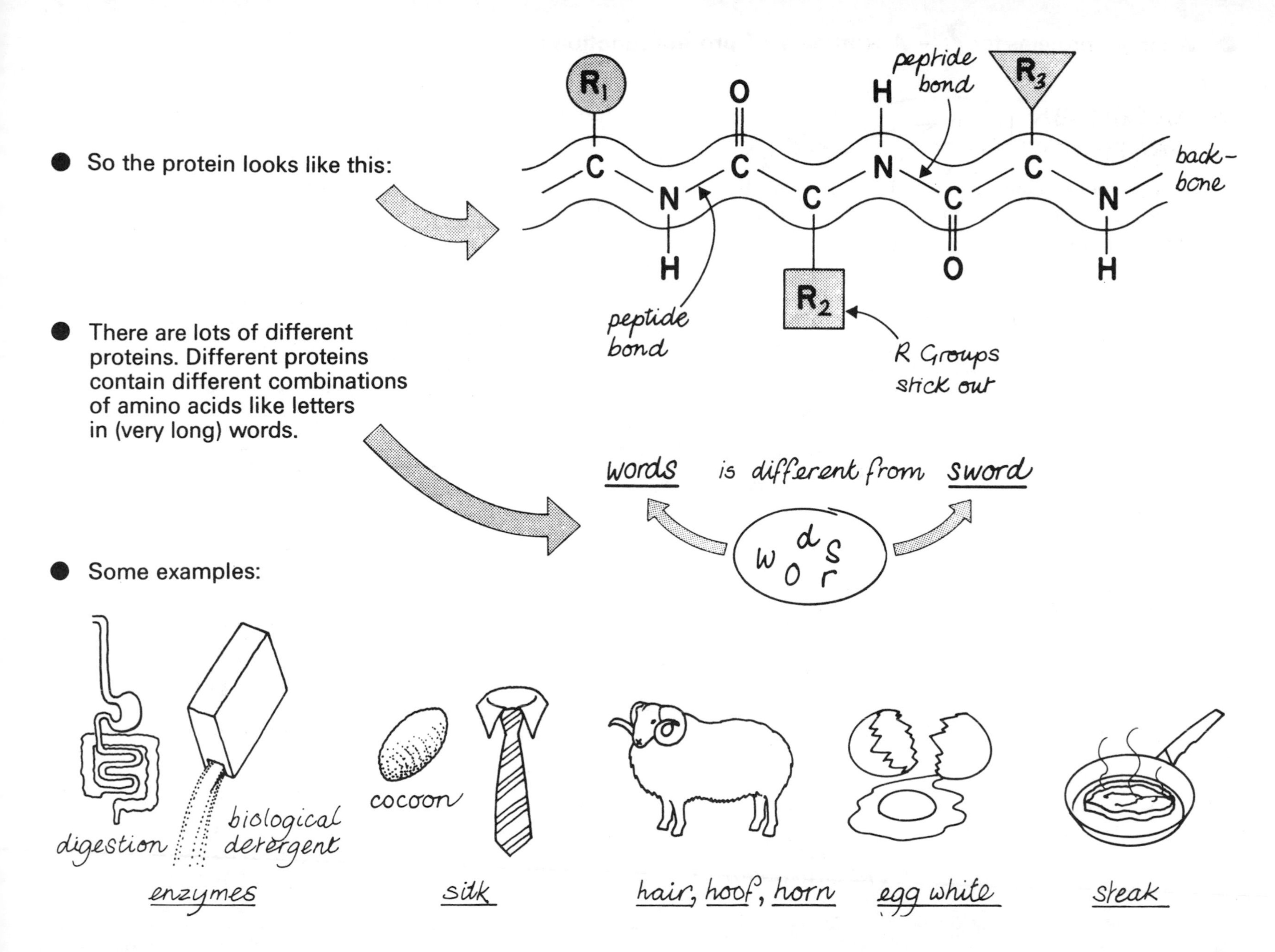

R1
O
H
peptide bond
R3
C
N
C
N
C
C
N
C
C
N
back-bone
H
R2
O
H
peptide bond
R Groups stick out
words is different from sword
w d o r s
digestion
biological detergent
enzymes
cocoon
silk
hair, hoof, horn
egg white
steak

● What are proteins for? – **A summary of protein functions**

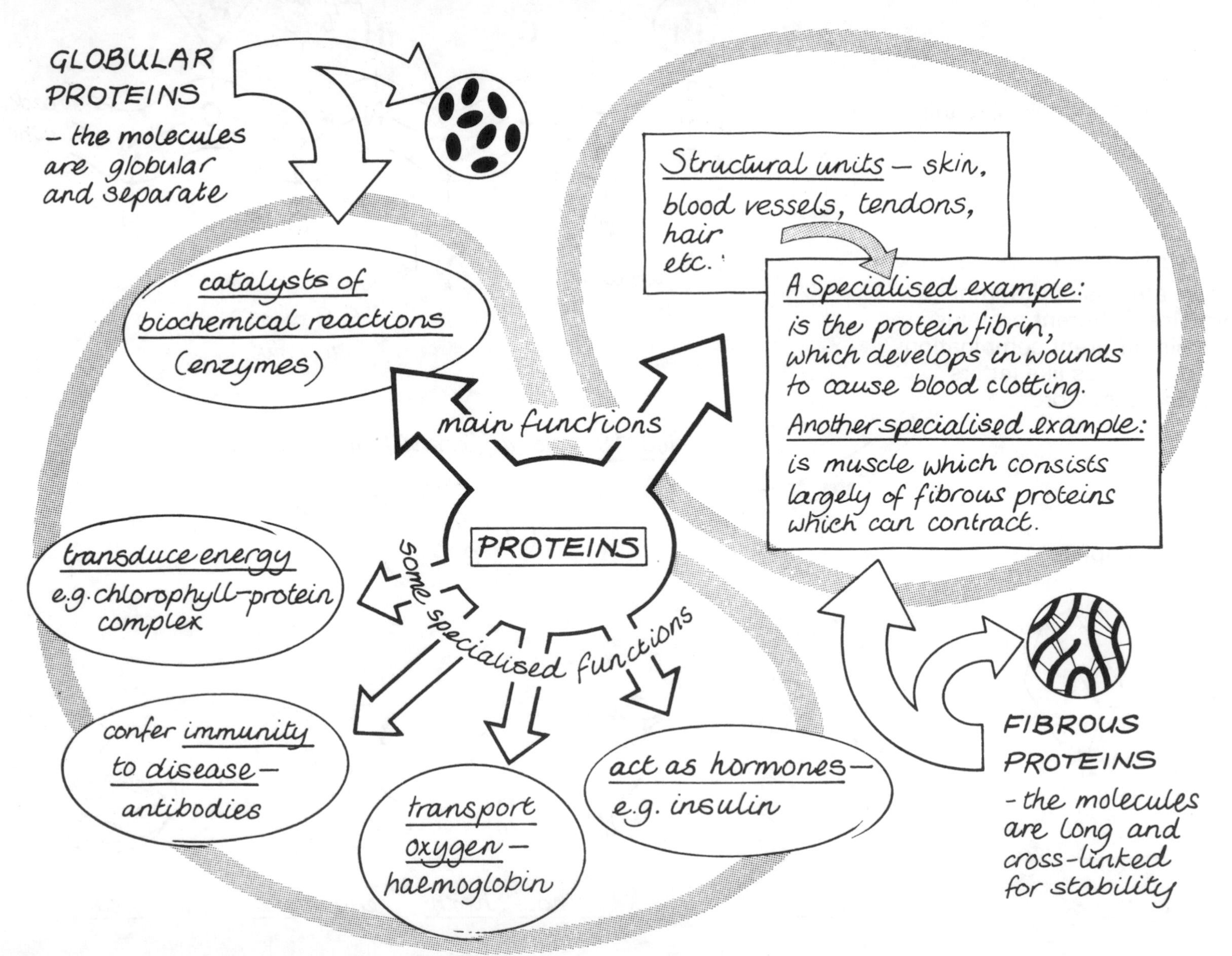

GLOBULAR PROTEINS
– the molecules are globular and separate
Structural units – skin, blood vessels, tendons, hair etc.
A Specialised example: is the protein fibrin, which develops in wounds to cause blood clotting.
Another specialised example: is muscle which consists largely of fibrous proteins which can contract.
catalysts of biochemical reactions (enzymes)
main functions
PROTEINS
some specialised functions
transduce energy e.g. chlorophyll-protein complex
confer immunity to disease – antibodies
transport oxygen – haemoglobin
act as hormones – e.g. insulin
FIBROUS PROTEINS
– the molecules are long and cross-linked for stability

- Proteins vary greatly in molecular weight

insulin
has 51 amino acids
(molecular weight 6000)

human haemoglobin
has 578 amino acids
(MW 63000)

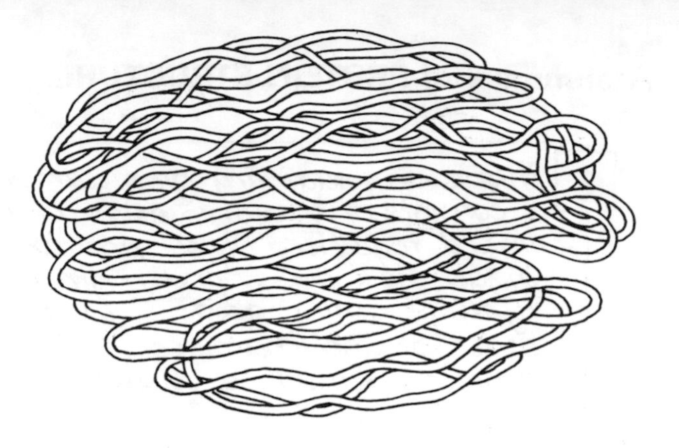

an enzyme (urease) from bean
has 4500 amino acids
(MW 473 000)

- Proteins are not just strings of amino acids. They have a definite shape.

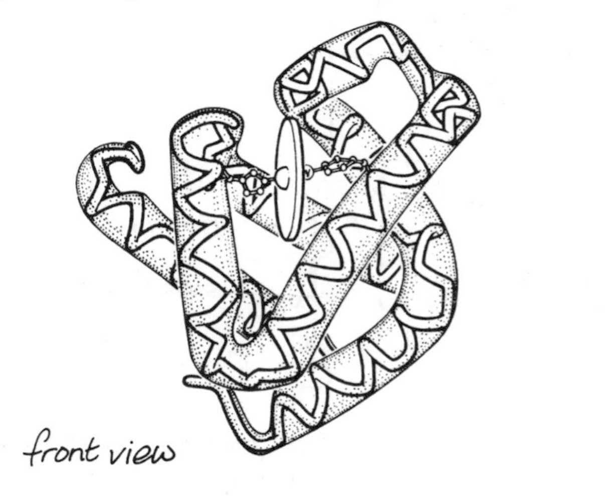

front view

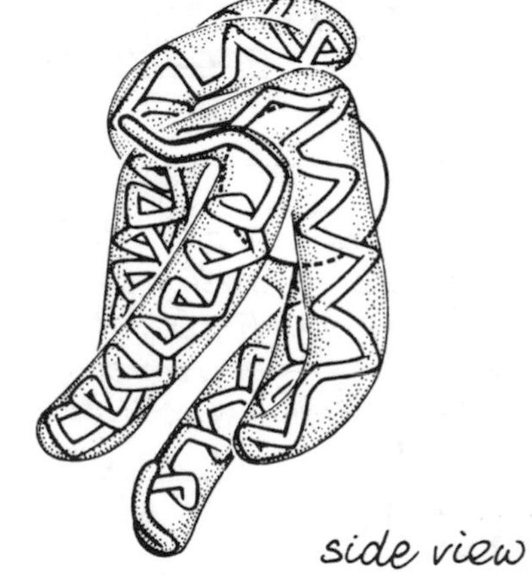

side view

Myoglobin molecule

A summary of PROTEIN STRUCTURE

1 This is an imaginary globular protein molecule. It consists of 77 amino acids (different types) linked together. (Most proteins are much bigger than this, and contain 20 different types of amino acids, but we can show most of the characteristics of typical proteins with this diagram.)

2 This protein molecule consists of 2 polypeptide chains A and B. Some proteins have only one chain. Some have several.

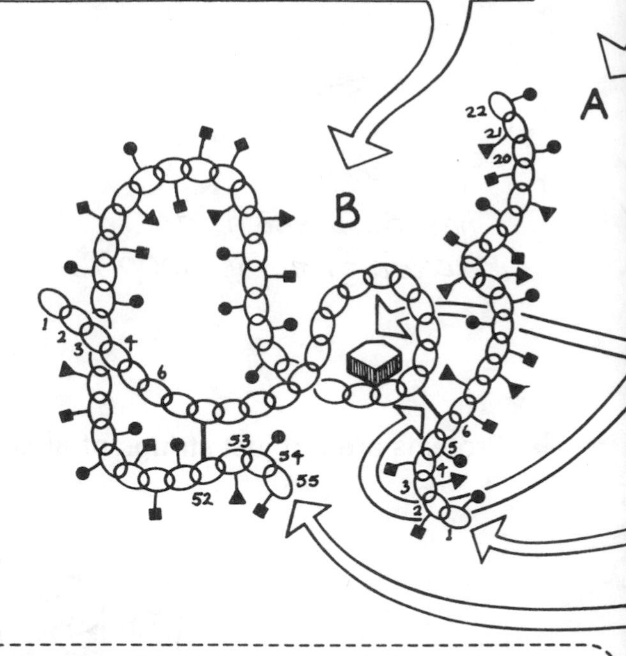

A
B

9 The whole 3-dimensional structure is called the tertiary structure of the protein. The chains do not 'flap' about at random. The tertiary structure is preset by interactions among the side groups of the amino acids. So each molecule of a particular protein has the same tertiary structure as every other molecule of that protein. If the 'normal' tertiary structure is destroyed e.g. by heat (boiling), or by strong acids or alkali or other reactive chemicals, the protein is said to be denatured. Denatured proteins often coagulate i.e. become insoluble.

8 The polypeptide chain is not a flat ribbon. For parts of it the amino acids are spirally arranged. This spiral is called the α-helix. The 3-dimensional structure of the 'ribbon' is called the secondary structure of the protein.

the secondary structure here is spiral

here it is not

diagram of protein with spiral and non-spiral parts to the chain

If = serine (ser) = glutamic acid (glu)
= lysine (lys) = glycine (gly)
= tryptophan (try)

The primary structure of A is:
Ser – lys – try – glu – ser – cys (teine) – lys – lys – gly – try – ser – ser – lys – gly – gly – glu – ser – gly – lys – ser – gly – ser

3 The chains are linked together by a sulphur bridge. This comes about when 2 S-containing amino acids (cysteine) link their S atoms together:

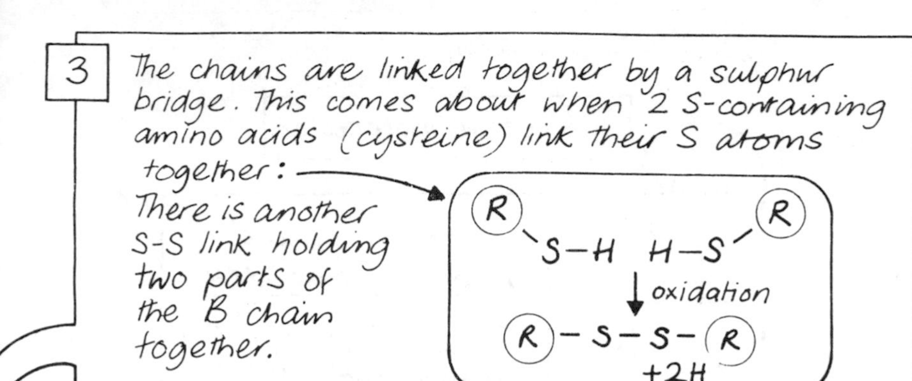
R
R
S–H H–S
oxidation
R – S – S – R
+2H

There is another S-S link holding two parts of the B chain together.

4 This is not an amino acid. It is a metal ion. Some proteins have other organic structures attached to them (often 'inside' like this one). Proteins which contain a structure which is not an amino acid are called conjugated proteins.

5 This is called the N-terminal end of the A-chain as it has a $-\overset{+}{N}H_3$ group which has not linked with another amino acid. By convention, the amino acids are counted from this end e.g. 'amino acid 6 of the A chain is cysteine.'

6 This is called the C-terminal end of the B-chain as it has a $-COO^-$ group which has not linked with another amino acid. The amino acids are counted from the other end e.g. 'amino acid 52 of the B-chain is cysteine.'

The other end of the B-chain must be N-terminal!

7 However much each chain is twisted in 3-dimensions, the amino acids have a particular sequence for that chain. This sequence is called the primary structure of the protein. It differs for different proteins.

CARBOHYDRATES

- Sucrose (cane or beet sugar) is an example of a carbohydrate.

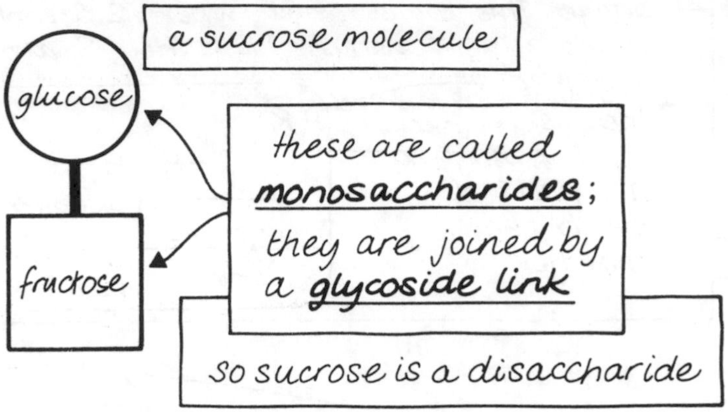

a sucrose molecule
glucose
fructose
these are called **monosaccharides**; they are joined by a **glycoside link**
so sucrose is a disaccharide

The words glyco- and sacchar- mean 'sugar'. So does the ending -ose.

- Lactose (milk sugar) is also a disaccharide.

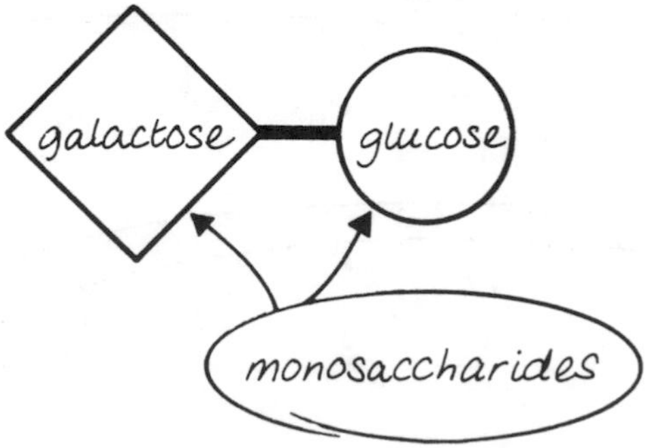

galactose
glucose
monosaccharides

- Starch is another carbohydrate. Each molecule has chains of hundreds or thousands of monosaccharide molecules joined together.

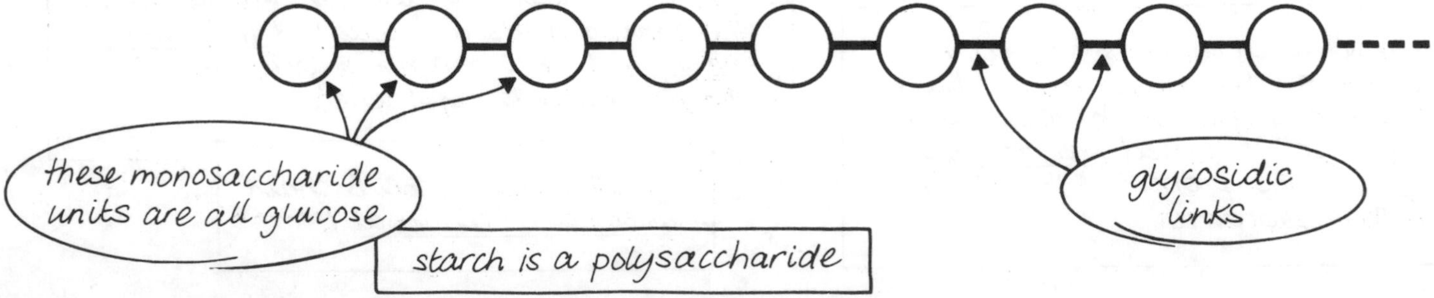

these monosaccharide units are all glucose
glycosidic links
starch is a polysaccharide

- Polysaccharide molecules may be branched. Any sample of starch contains two sorts of molecule – one unbranched, the other branched.

These molecules take up shapes in 3 dimensions.

About 20% of starch is amylose.
About 80% of starch is amylopectin.

- Glycosidic links can be broken, releasing the monosaccharide sugars.

- Some monosaccharides are found free in nature – e.g. glucose in blood, fructose in fruits.

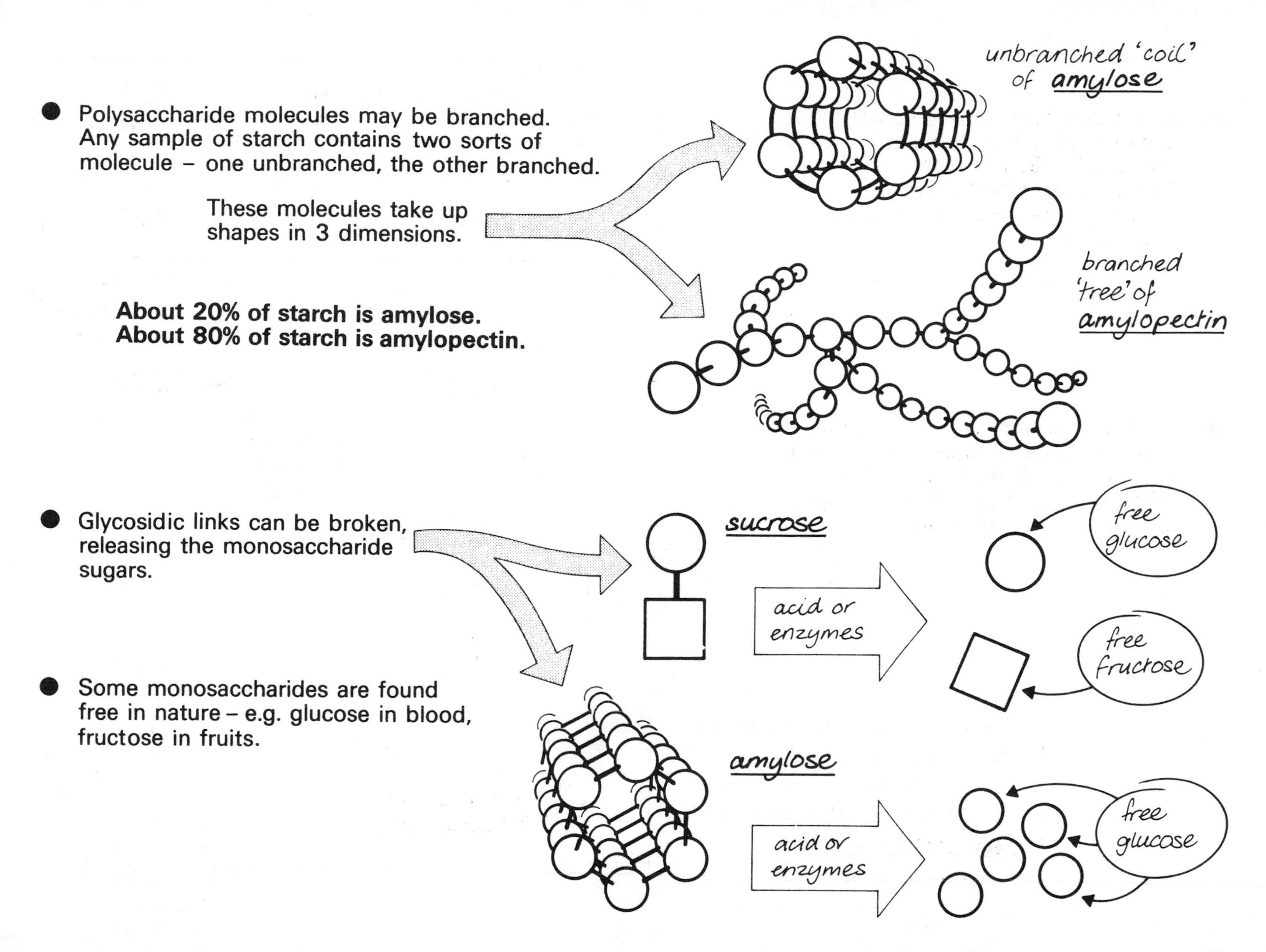
unbranched 'coil' of amylose
branched 'tree' of amylopectin
sucrose
acid or enzymes
free glucose
free fructose
amylose
acid or enzymes
free glucose

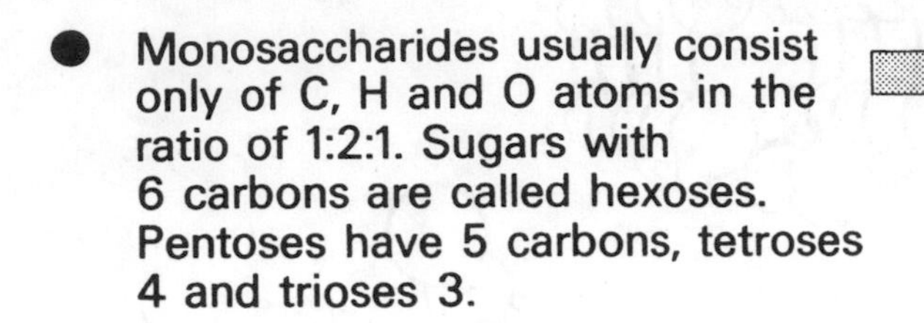

- Monosaccharides usually consist only of C, H and O atoms in the ratio of 1:2:1. Sugars with 6 carbons are called hexoses. Pentoses have 5 carbons, tetroses 4 and trioses 3.

glucose (a hexose)	$C_6H_{12}O_6$
xylose (a pentose)	$C_5H_{10}O_5$
erythrose (a tetrose)	$C_4H_8O_4$
glyceraldehyde (a triose)	$C_3H_6O_3$

trioses are important in metabolism

xylose is found as a polysaccharide in wood

- Molecules of different sugars often have the same number of atoms but they are joined together in slightly different ways. The formula $C_3H_6O_3$ represents more than one sugar.

glyceraldehyde

this is an aldehyde group

a different sugar- dihydroxyacetone

this is a keto group

Aldehyde and keto groups can **chemically reduce** other compounds. So these monosaccharides are called **reducing sugars**

- Glyceraldehyde itself can exist as 2 enantiomers – 'mirror images'.

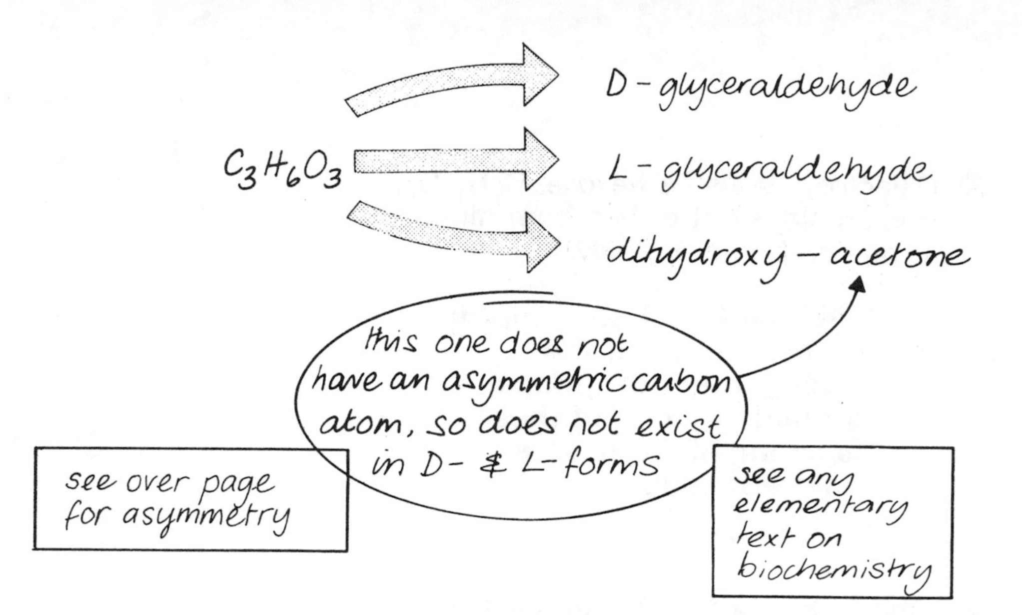

$C_3H_6O_3$
D-glyceraldehyde
L-glyceraldehyde
dihydroxy-acetone
this one does not have an asymmetric carbon atom, so does not exist in D- & L-forms
see over page for asymmetry
see any elementary text on biochemistry

- So the formula $C_3H_6O_3$ represents 3 different sugars.

- Naturally-occurring sugars are nearly all D-sugars.

- Glucose and fructose are **reducing sugars** They are both hexoses, $C_6H_{12}O_6$. Glucose is an aldose (it has an aldehyde group) Fructose is a ketose (it has a keto group).

for convenience the carbons are numbered like this

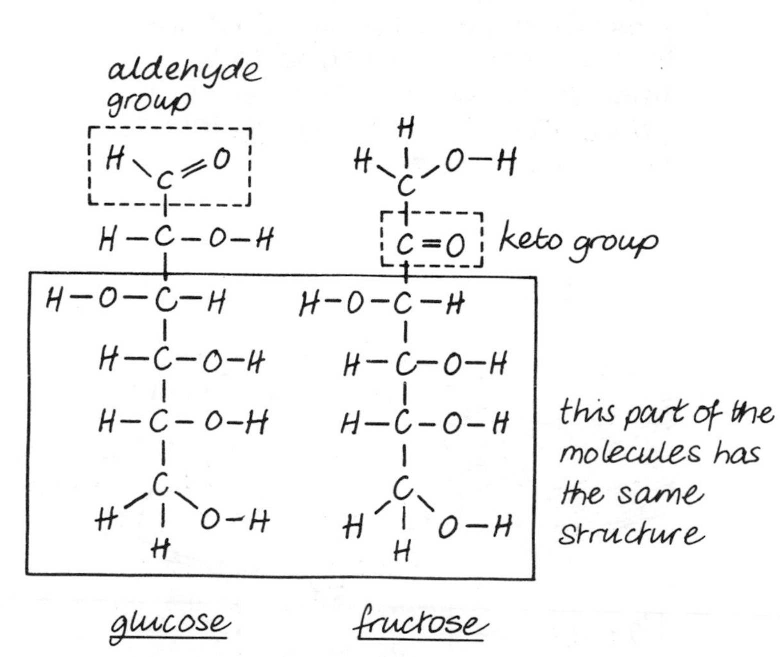

aldehyde group
keto group
1
2
3
4
5
6
this part of the molecules has the same structure
glucose
fructose

- Galactose is also a hexose, $C_6H_{12}O_6$. It is an aldose. It differs from glucose at one carbon atom only.

  Glucose has 4 asymmetric carbon atoms. If the arrangement of groups around any one of these is changed we have a different sugar.

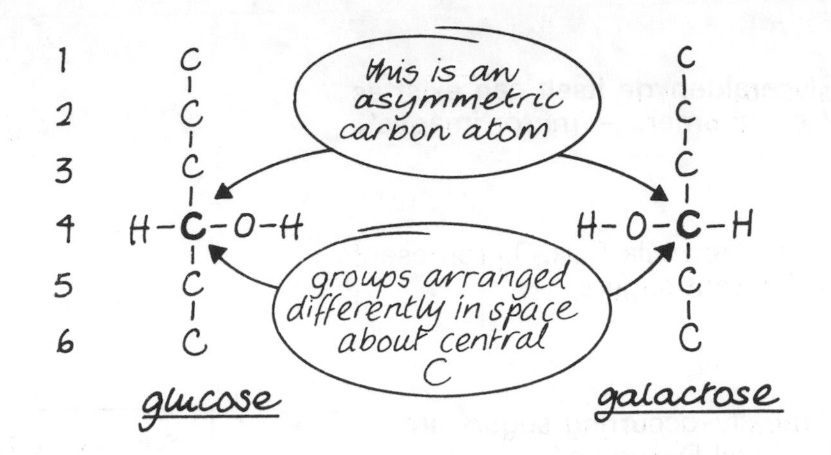

1
2
3
4
5
6
C
C
C
H–C–O–H
C
C
this is an asymmetric carbon atom
groups arranged differently in space about central C
C
C
C
H–O–C–H
C
C
glucose
galactose

- There are about 20 naturally occurring types of monosaccharide. Most are hexoses (e.g. glucose, galactose, fructose) or pentoses (e.g. xylose, ribose). All the carbohydrates found in living organisms are made from them.

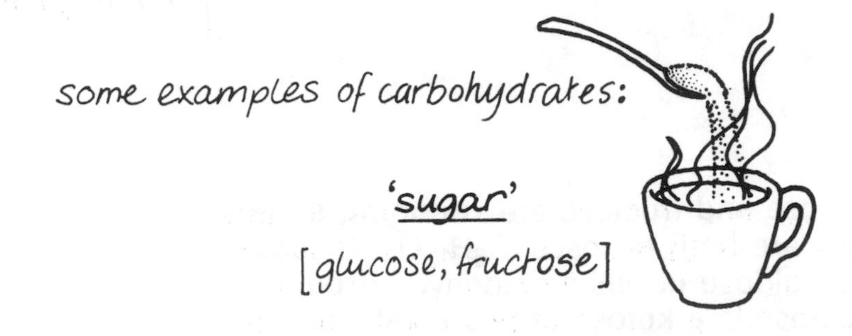

some examples of carbohydrates:
'sugar'
[glucose, fructose]

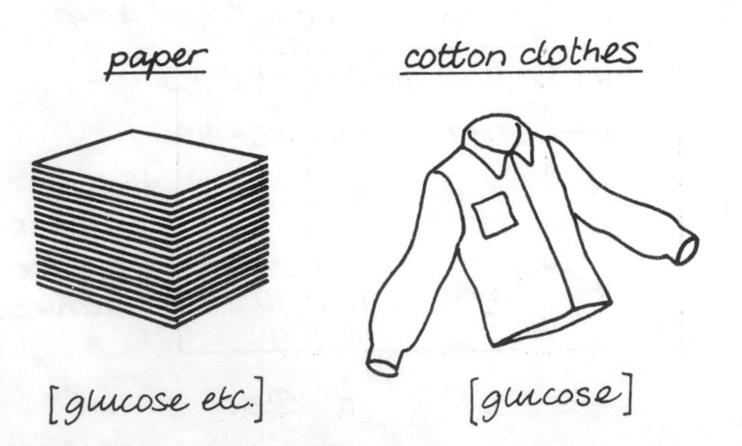

paper
cotton clothes
[glucose etc.]
[glucose]

pectin
(makes jam set)
Jam

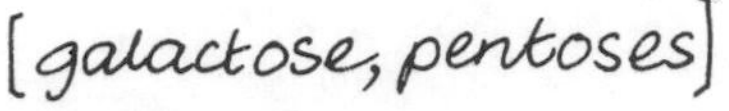

[galactose, pentoses]

wooden furniture
[xylose, glucose & others]

More about carbohydrate structure

- A glycosidic link is made when the reducing (aldehyde or keto) group of a monosaccharide reacts EITHER with an alcoholic (hydroxyl) group of another molecule – for instance another sugar molecule OR – less commonly – with the reducing group of another sugar.

- The chain of carbon atoms for a monosaccharide is not straight in 3 dimensions – we show it that way for convenience. It looks more like this.

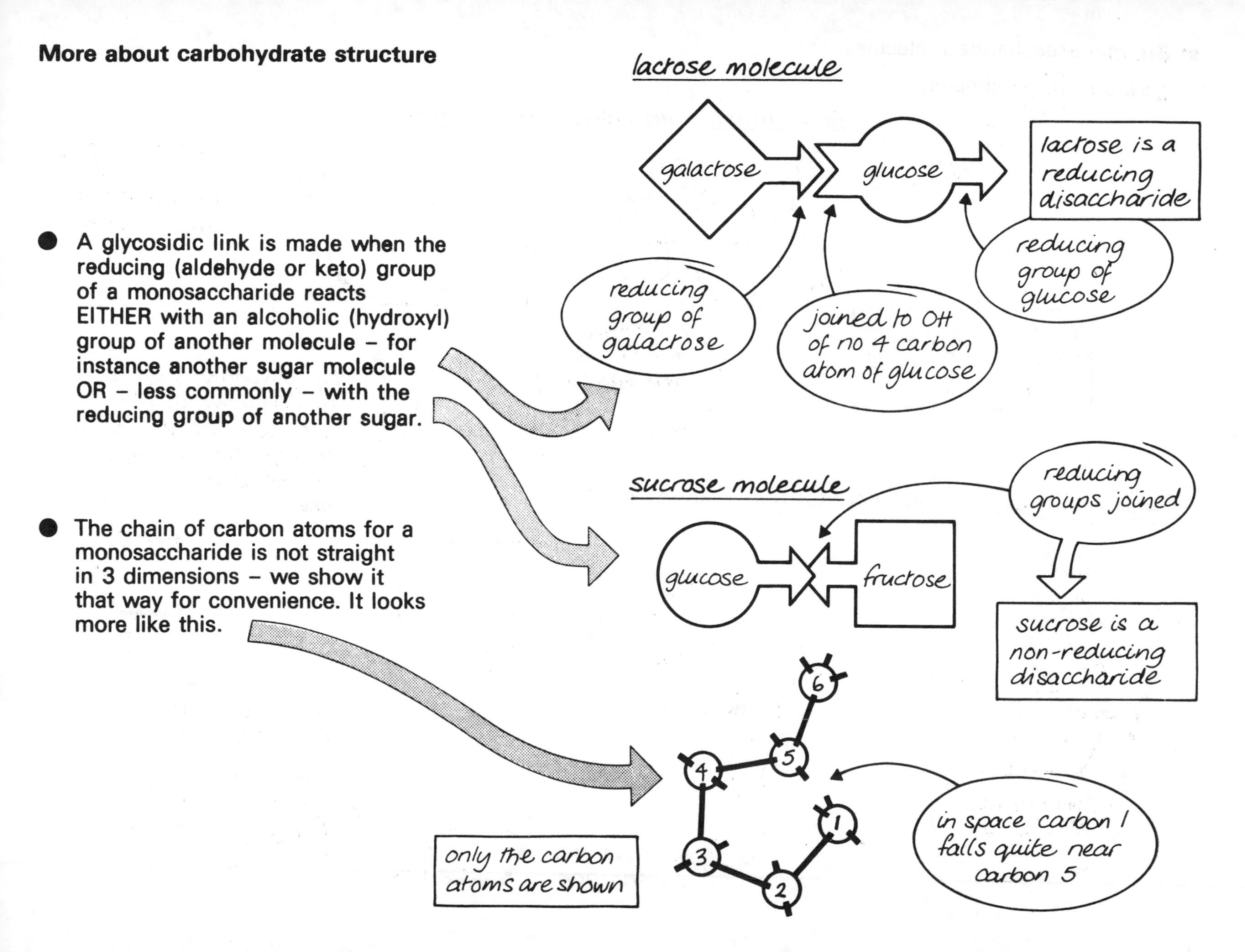

lactose molecule
galactose
glucose
lactose is a reducing disaccharide
reducing group of galactose
joined to OH of no 4 carbon atom of glucose
reducing group of glucose
sucrose molecule
reducing groups joined
glucose
fructose
sucrose is a non-reducing disaccharide
6
5
4
3
2
1
in space carbon 1 falls quite near carbon 5
only the carbon atoms are shown

● **So, monosaccharide molecules take up ring shapes:**

for example – an aldohexose (e.g. glucose)

6C, 5C, 4C, 3, 2, 1, O, H, C

becomes

this is still the position of the reducing group

for simplicity, most of the H & O atoms are omitted

this O links very easily to C1

by convention this is written down as if it were a hexagon looked at from the viewpoint of the arrow – like this:

another example – a ketohexose (e.g. fructose)

becomes

C2 is still the position of the reducing group

this is called a **pyranose** ring

this is called a **furanose** ring

this O links very easily to C2

by convention a pentagon

- In naturally occurring di-, tri-, and polysaccharides, glycosidic links join rings like these together – for example **amylose**.

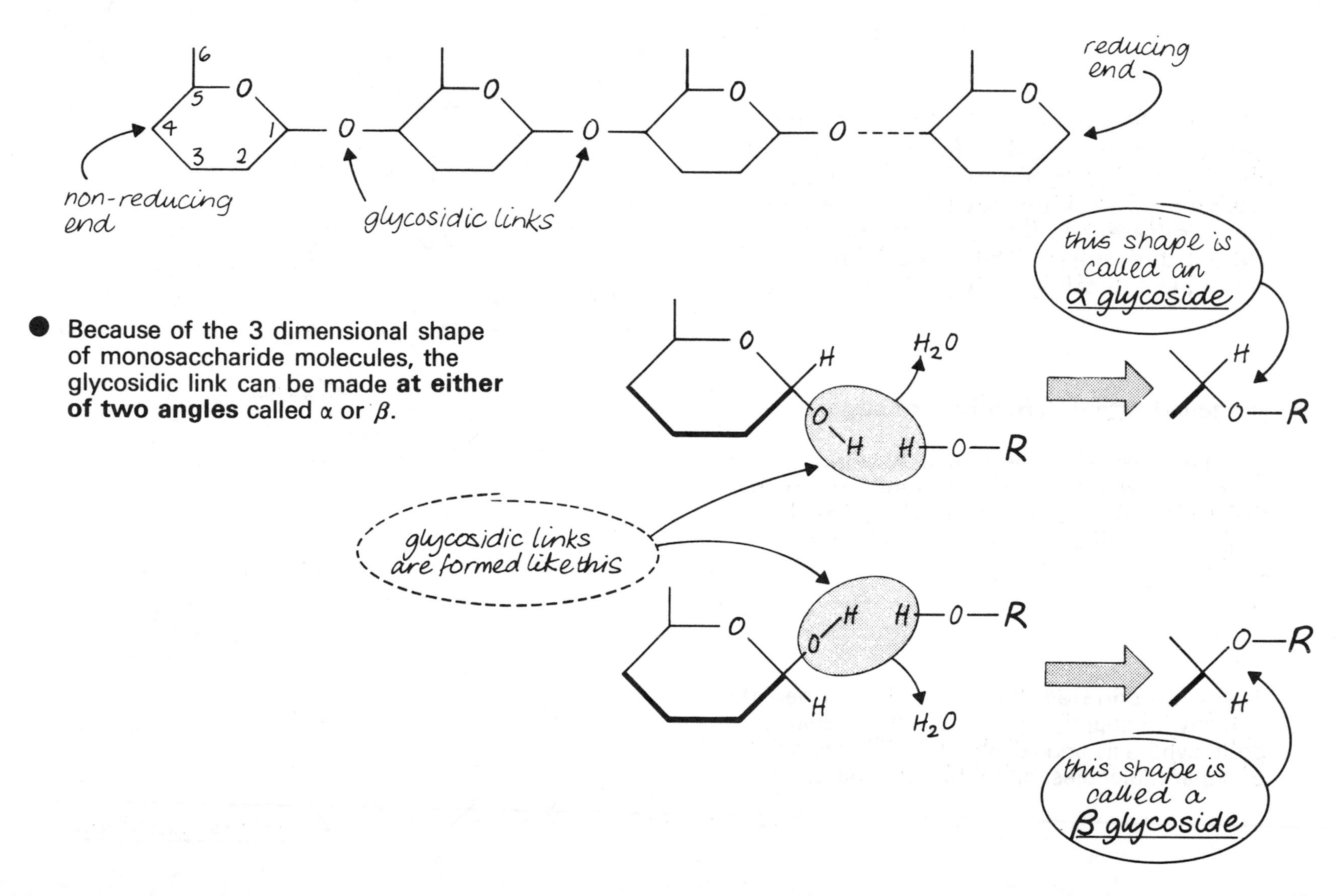

6
5
4
3
2
1
O
reducing end
non-reducing end
glycosidic links
this shape is called an α glycoside
H_2O
H
O—H
H—O—R
glycosidic links are formed like this
this shape is called a β glycoside

- Because of the 3 dimensional shape of monosaccharide molecules, the glycosidic link can be made **at either of two angles** called α or β.

- These angles give rise to chains of completely different shape. The concept can be demonstrated like this: take building blocks.

α angle gives coiled structure (e.g. amylose) leading to granular material (starch).

α angle

β

α

slot at fixed angle

tongue at angle α or β

The only difference between starch and cellulose is the angle of the glycoside linkages between the monosaccharide (glucose) molecules. Yet these molecules are so different in their properties that we eat one (bread, custard powder) and wear the other (cotton clothes).

β angle

β angle gives elongated molecules (e.g. cellulose) which can lie side by side to give fibres or sheets.

- **Some important derivatives of sugars**

Some monosaccharides contain N atoms in their molecules. They are called **amino sugars.** Polysaccharides can be formed from them (e.g. chitins found in insect exoskeletons).

Sugar phosphates are important intermediates in many biochemical processes (for instance photosynthesis and respiration). They consist of sugar molecules joined to phosphate ions.

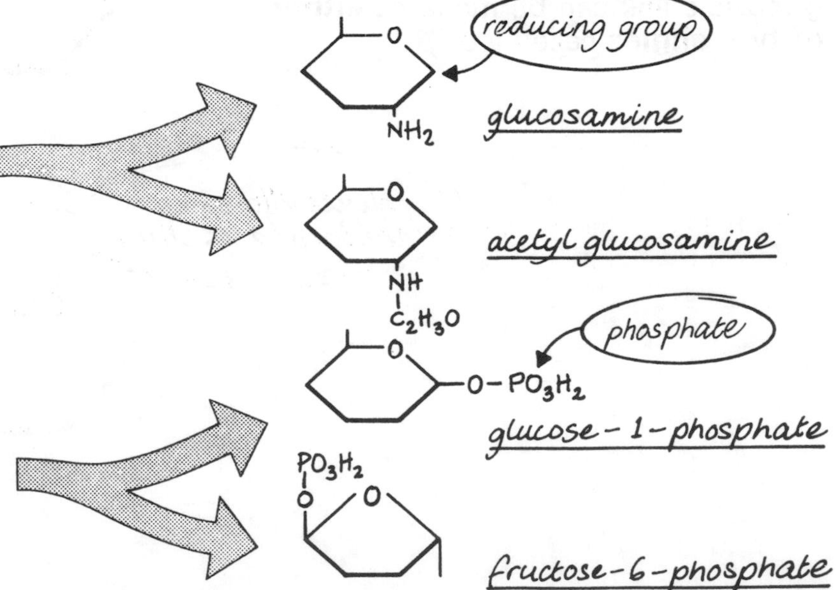

O
reducing group
NH2
glucosamine
O
acetyl glucosamine
NH
C2H3O
phosphate
O
O–PO3H2
glucose-1-phosphate
PO3H2
O
O
fructose-6-phosphate

FATS

- Fats (and oils derived from living things) are insoluble in water but soluble in organic solvents like acetone, ether and petrol.

- A fat (lard for instance) or oil (olive oil) is made up of compounds called **triglycerides.** Triglycerides consist only of C, H & O.

- Triglycerides are hydrolysed by acid, alkali or certain enzymes to **glycerol** and **fatty acids**.

- Glycerol has 3 alcohol groups in each molecule.

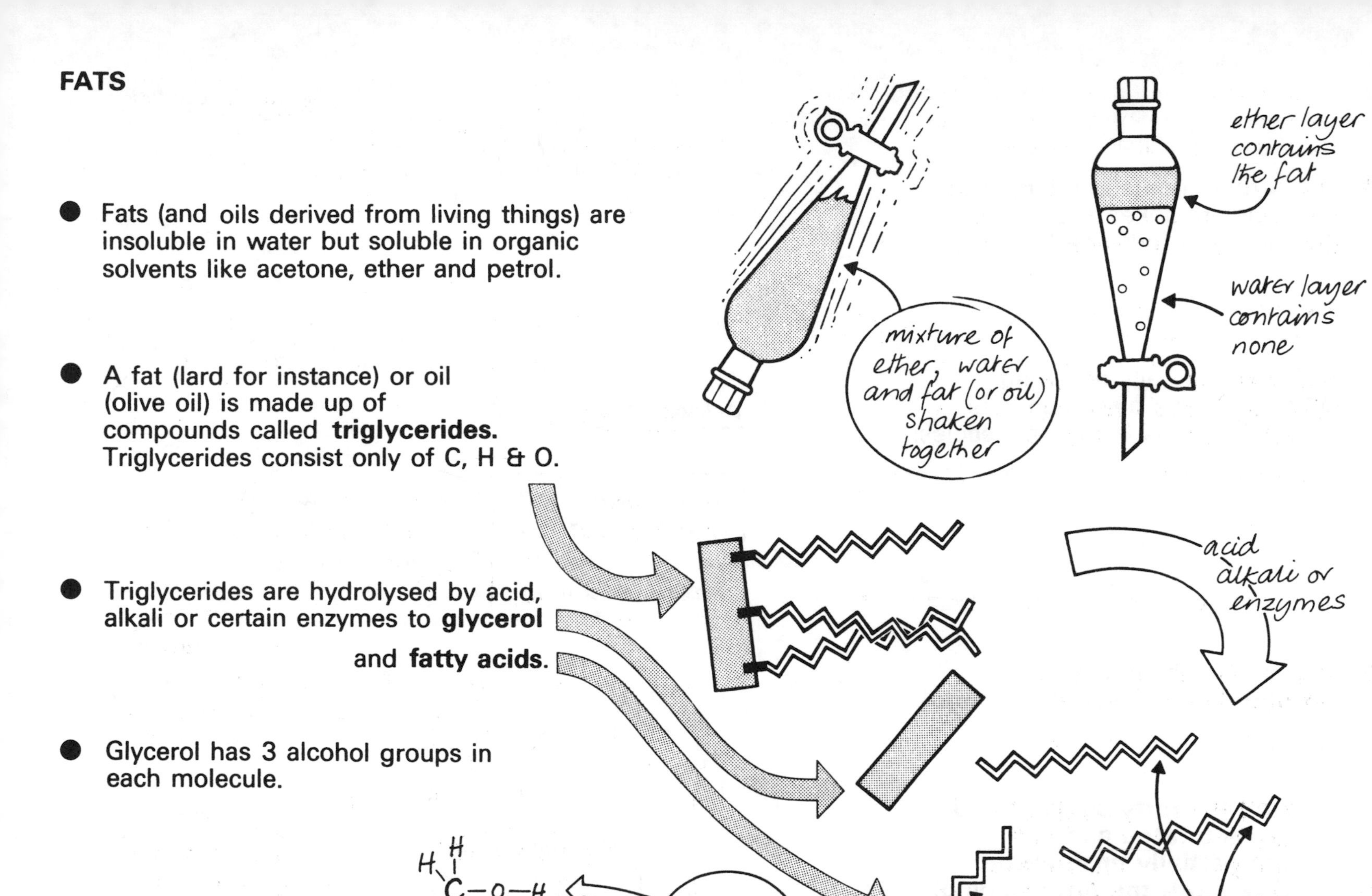

ether layer contains the fat
mixture of ether, water and fat (or oil) shaken together
water layer contains none
acid alkali or enzymes
these molecules need not all be exactly the same
alcohol groups
H
H
C—O—H
H—C—O—H
C—O—H
H
H

- The carbon atoms in a fatty acid molecule are linked together in the form of a **hydrocarbon** chain.

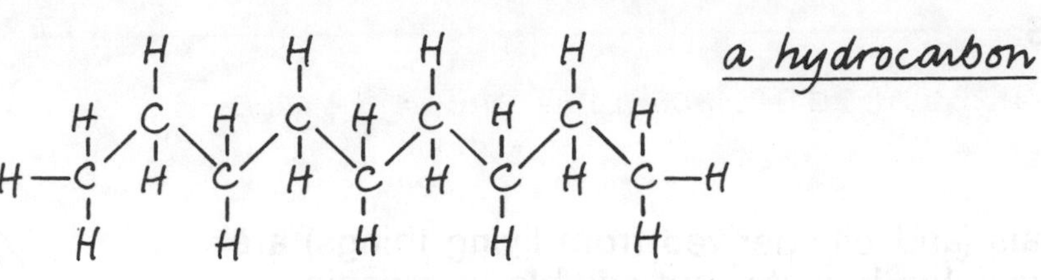
a hydrocarbon

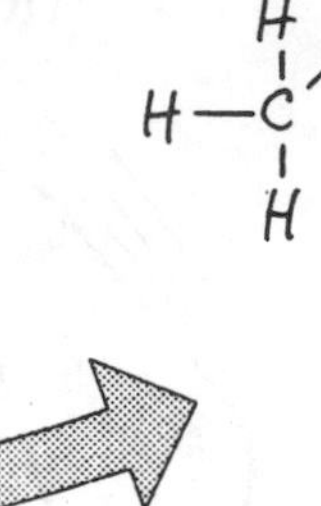

a fatty acid

- Nearly all naturally occurring fatty acids have **even** numbers of carbon atoms in their molecules.

lauric acid	12
myristic acid	14
palmitic acid	16
stearic acid	18

number of carbon atoms in each molecule

- The carbon chain may be **saturated** or partially **unsaturated**.

a saturated hydrocarbon chain

double bonds

note how an unsaturated hydrocarbon chain has a different shape

Half the fatty acids of lard are saturated and half are partially unsaturated
One-tenth the fatty acids of olive oil are saturated and nine-tenths partially unsaturated.

- Each fatty acid molecule has one acid group.

rest of molecule → (R)—C(=O)—O—H [boxed: acid group]; O—H ← replaceable H

- The fatty acid molecules are linked to glycerol molecules by **ester** links.

- Fatty acid molecules are much larger than glycerol molecules.

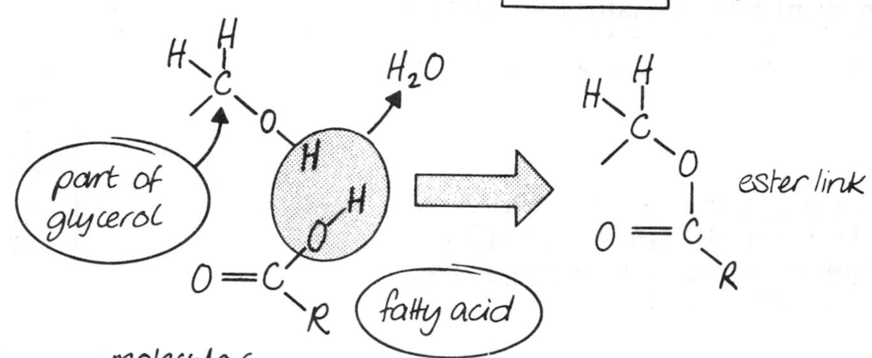

part of glycerol
H_2O
fatty acid
ester link

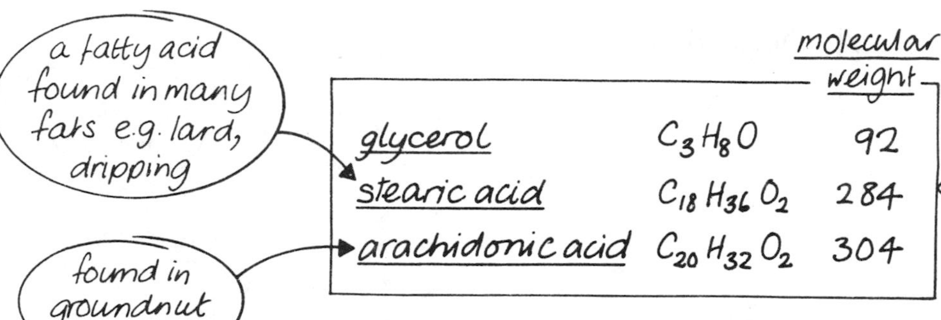

a fatty acid found in many fats e.g. lard, dripping
found in groundnut oil

		molecular weight
glycerol	C_3H_8O	92
stearic acid	$C_{18}H_{36}O_2$	284
arachidonic acid	$C_{20}H_{32}O_2$	304

As each molecule of fat consists of 1 molecule of glycerol and 3 molecules of fatty acid, most of the weight of a fat is fatty acid.

- Unsaturated fatty acids have a lower melting point than saturated fatty acids.

solid at room temperature

Stearic acid (C_{18}) has 0 double bonds – its melting point is 70°C

Oleic acid (C_{18} also) has 1 double bond – its melting point is 13–16°C

liquid at room temperature

– so are linoleic (C_{18}, 2 double bonds) and linolenic (C_{18}, 3 double bonds)

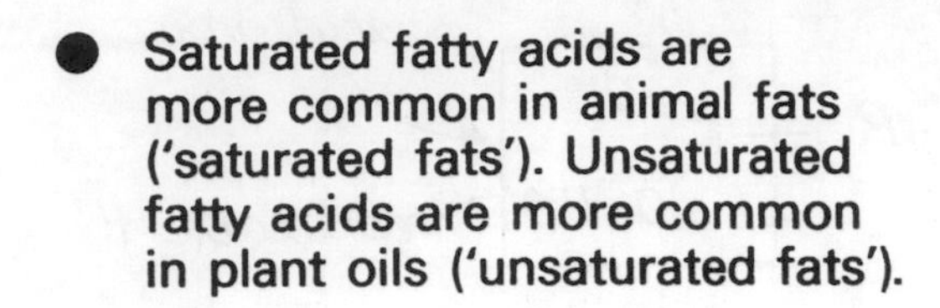

- Saturated fatty acids are more common in animal fats ('saturated fats'). Unsaturated fatty acids are more common in plant oils ('unsaturated fats').

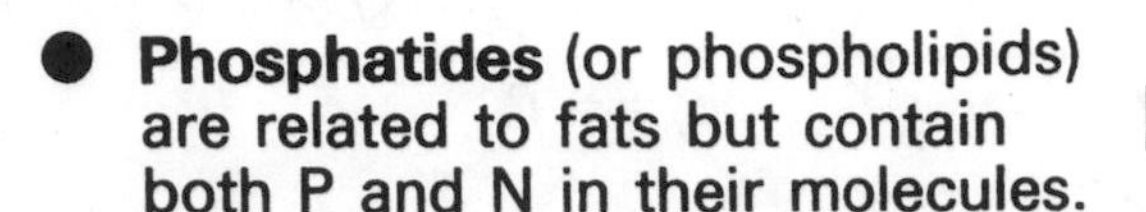

- **Phosphatides** (or phospholipids) are related to fats but contain both P and N in their molecules.

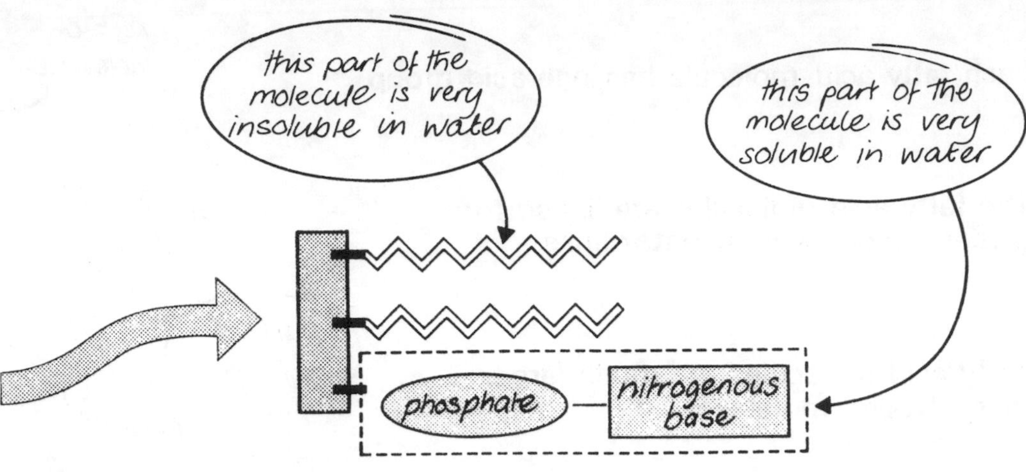
this part of the molecule is very insoluble in water
this part of the molecule is very soluble in water
phosphate
nitrogenous base

A phosphatide

- Because of the great difference in solubility of the two parts of phosphatide molecules, they take up a layered position at a water surface, like this:

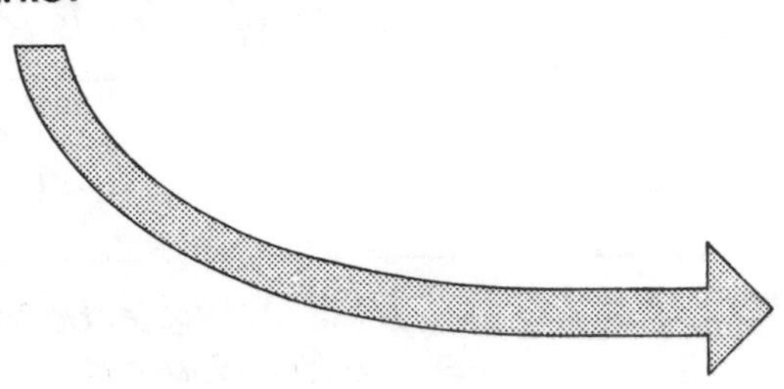

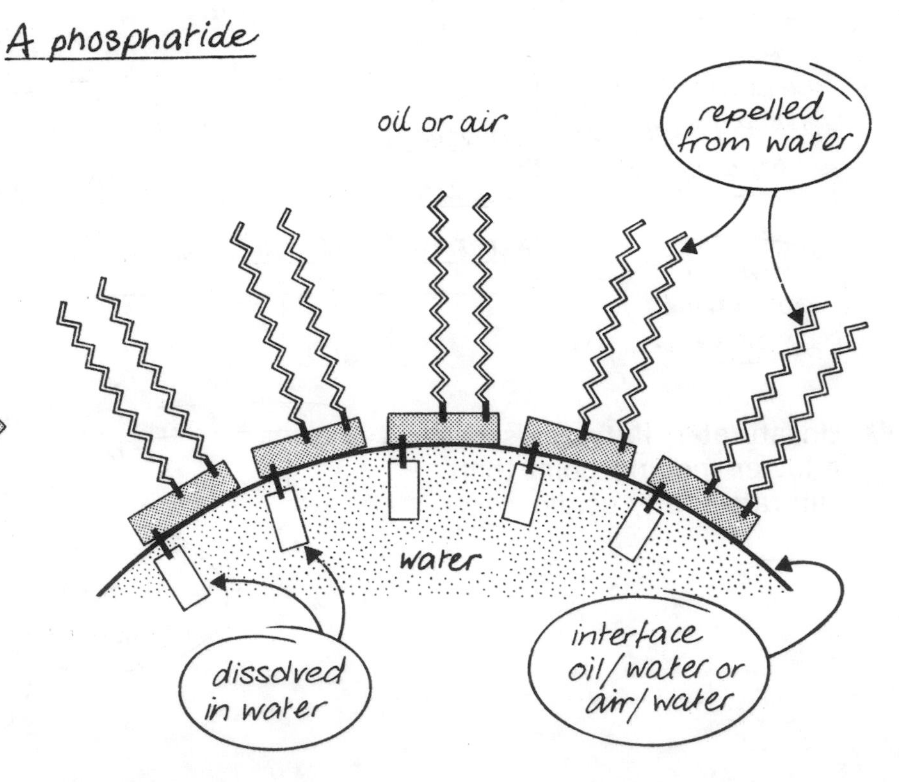
oil or air
repelled from water
water
dissolved in water
interface oil/water or air/water

– and so help to make a membrane. They are found as important constituents of subcellular membranes (of mitochondria, endoplasmic reticulum, for example) of all living things.

NUCLEIC ACIDS

- There are two fundamental types of nucleic acid: **ribonucleic acid** and **deoxyribonucleic acid**.
- They are called **RNA** and **DNA**.
- Nucleic acids are important because

RNA 'makes' all the proteins found in cells.

DNA carries the 'codes' in the cell nuclei which pass on the information about the types of protein to be made.

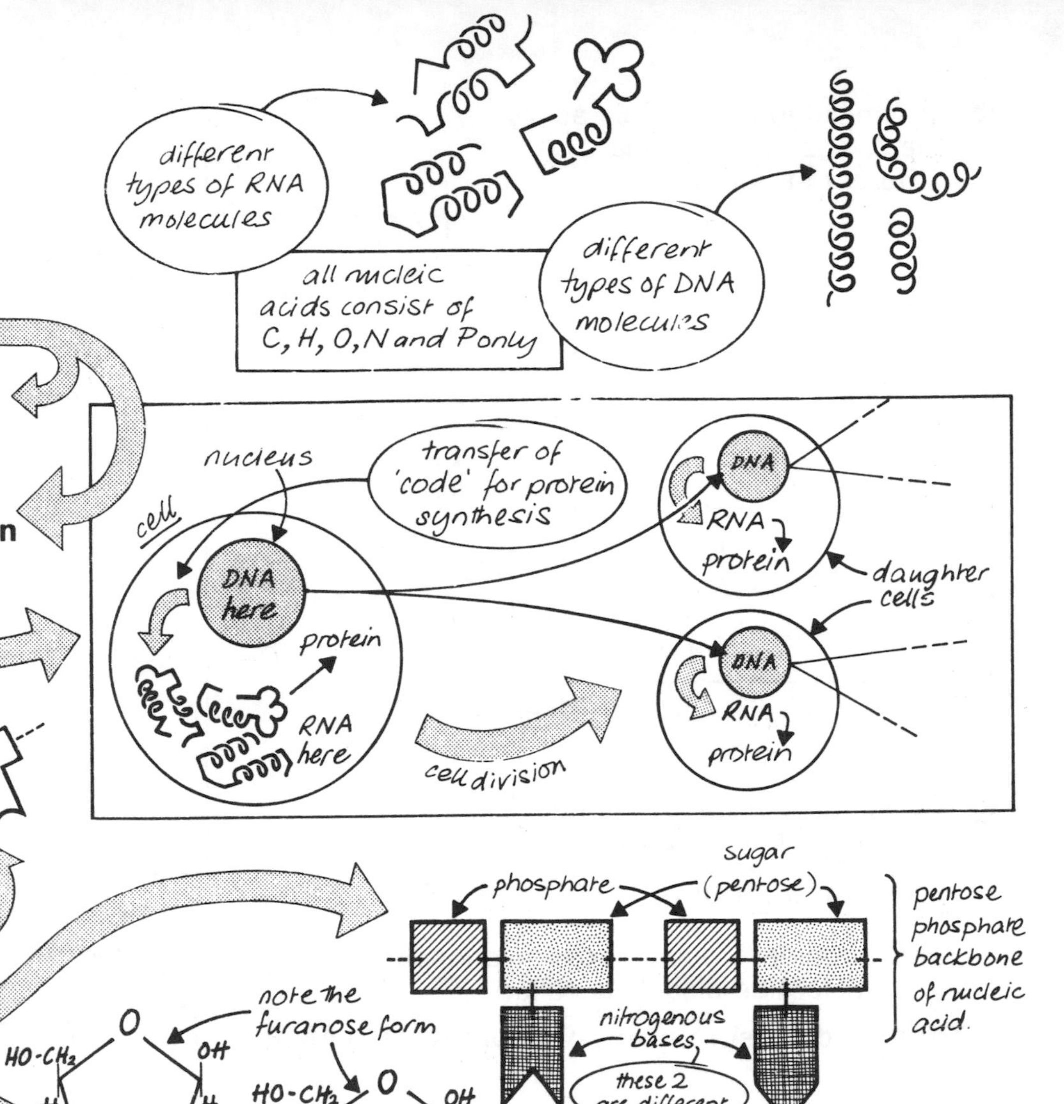

different types of RNA molecules
all nucleic acids consist of C, H, O, N and P only
different types of DNA molecules
nucleus
transfer of 'code' for protein synthesis
cell
DNA here
protein
RNA here
cell division
DNA
RNA
protein
daughter cells
sugar
phosphate
(pentose)
pentose phosphate backbone of nucleic acid
nitrogenous bases
these 2 are different
note the furanose form
O
HO-CH₂
OH
H

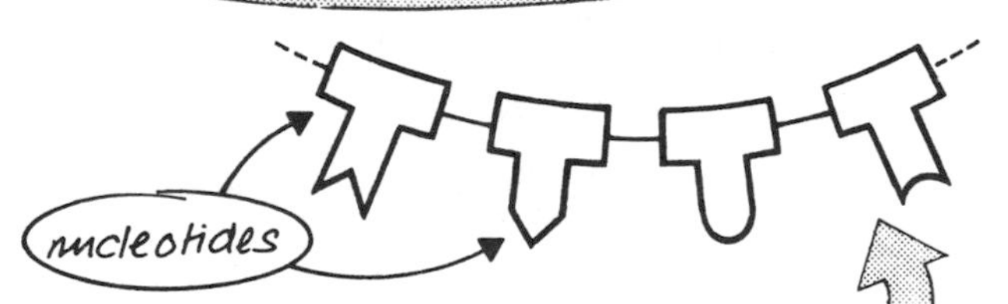

nucleotides

- Nucleic acids consist of chains of **nucleotides** joined together.
- A nucleotide is itself a 'packet' of 3 molecules joined together.
- In RNA the sugar in the backbone is **ribose**.
- In DNA the sugar in the backbone is **deoxyribose**.

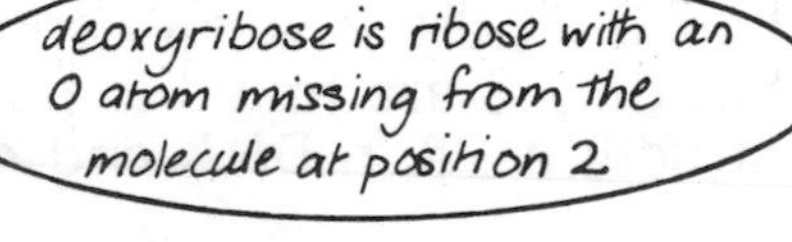

deoxyribose is ribose with an O atom missing from the molecule at position 2

- In either type of nucleic acid, the N-base can be one of 4 different structures.

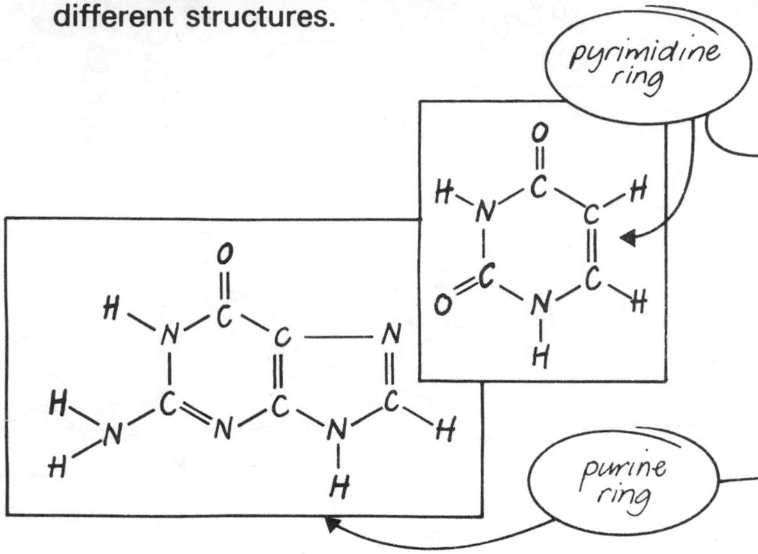

pyrimidine ring

purine ring

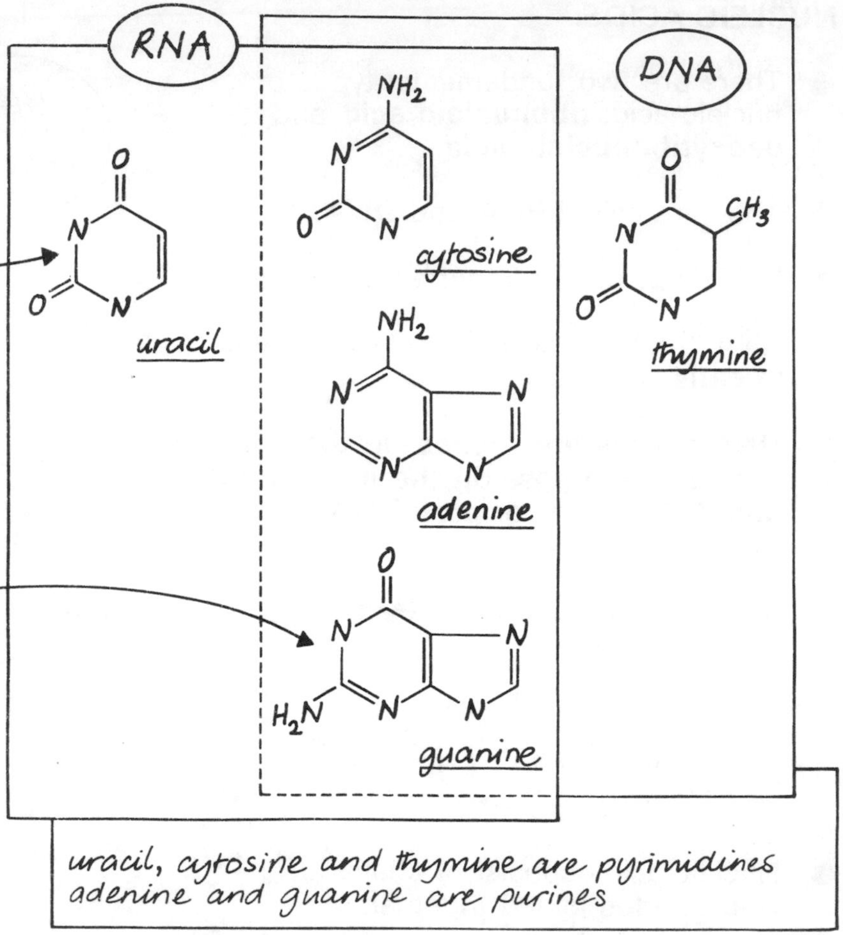

RNA

DNA

uracil

cytosine

adenine

guanine

thymine

uracil, cytosine and thymine are pyrimidines
adenine and guanine are purines

- **So** adenine – ribose – phosphate is a nucleotide. It is called

adenosine monophosphate or AMP

and adenine – deoxyribose – phosphate is a nucleotide also. It is called

deoxyadenosine monophosphate or dAMP.

	adenine A	guanine G	cytosine C	uracil U	thymine T
RNA	AMP	GMP	CMP	UMP	not found
DNA	dAMP	dGMP	dCMP	not found	dTMP

- There would be 10 possible nucleotides based on the 2 sugars and 5 bases, but only 8 are found in nucleic acid structures, 4 in DNA and 4 in RNA.

- **Remember:** DNA differs chemically from RNA because:

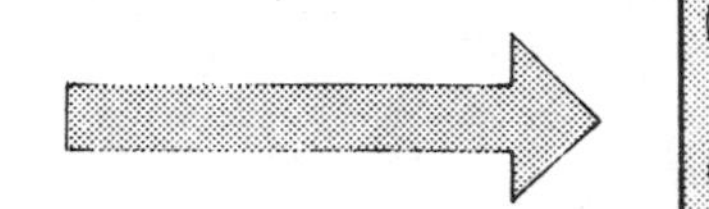

1. deoxyribose is the sugar found in the backbone
2. uracil is <u>not</u> one of the bases
3. thymine <u>is</u> one of the bases

- and therefore we can represent

RNA like this:

and DNA like this:

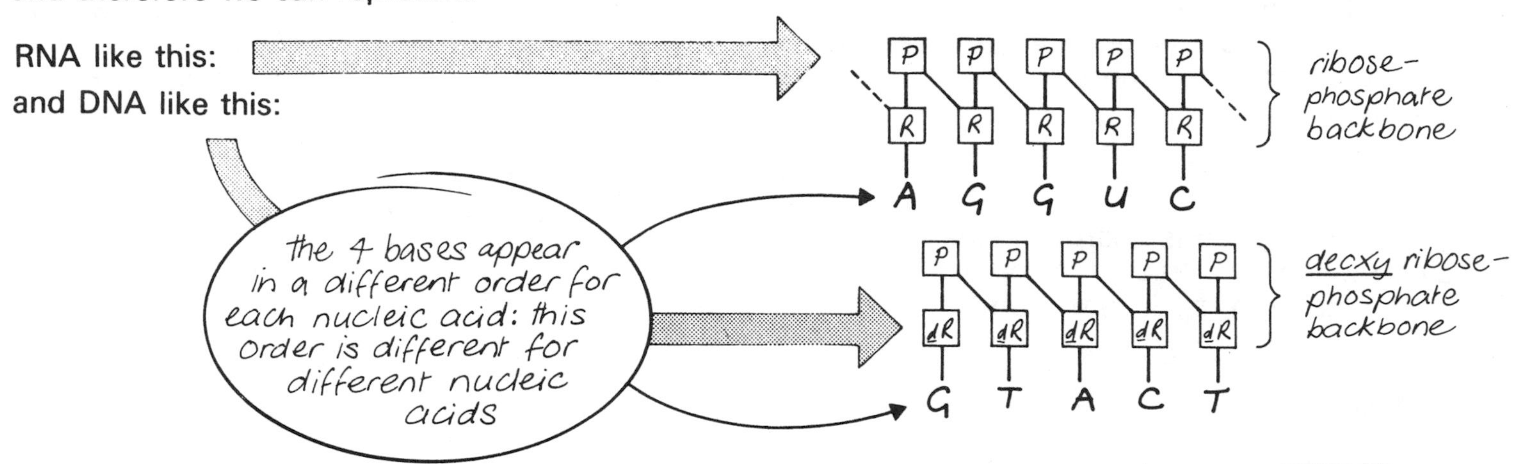

P P P P P
R R R R R
A G G U C
ribose-phosphate backbone
the 4 bases appear in a different order for each nucleic acid: this order is different for different nucleic acids
P P P P P
dR dR dR dR dR
G T A C T
deoxy ribose-phosphate backbone

- **The shape of RNA and DNA molecules in space is important.**

- DNA molecules occur in the cell nucleus in the shape of a **double helix** (spiral). This shape is taken up because the bases attract one another by **hydrogen bonding** forming parallel rungs of a spiral ladder.

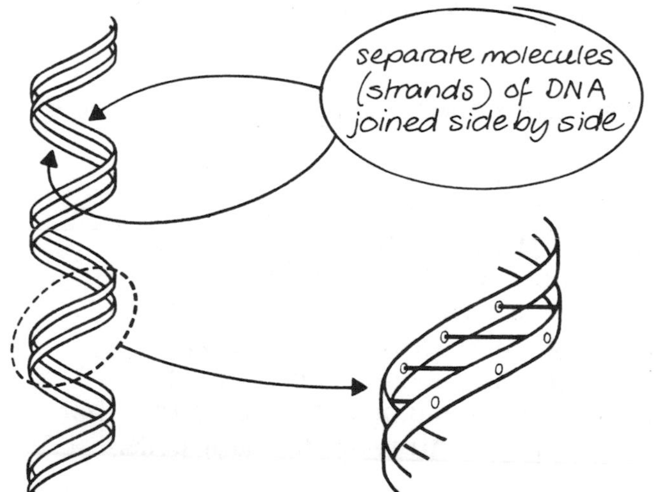

separate molecules (strands) of DNA joined side by side

A diversion

HYDROGEN BONDING

Consider an **—OH** group. It is **electronically neutral** as a whole, but if we look within the group, a part of it will be more negatively charged, and another part more positively charged.

electronically neutral as a whole

R—O H

this part of the group contains the 'extra' bonding electron (from H) and is hence -ve

this part of the group lacks the bonding electron and is hence +ve

If another similar group is nearby, a weak link may be made by attraction of opposite charges. It is called a **hydrogen bond**.

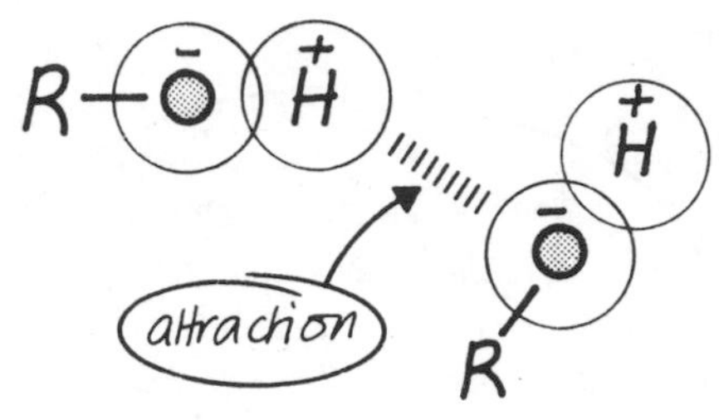

attraction

TWO EXAMPLES:

Peptide groups readily form H-bonds.

Nucleotide bases readily form H-bonds.

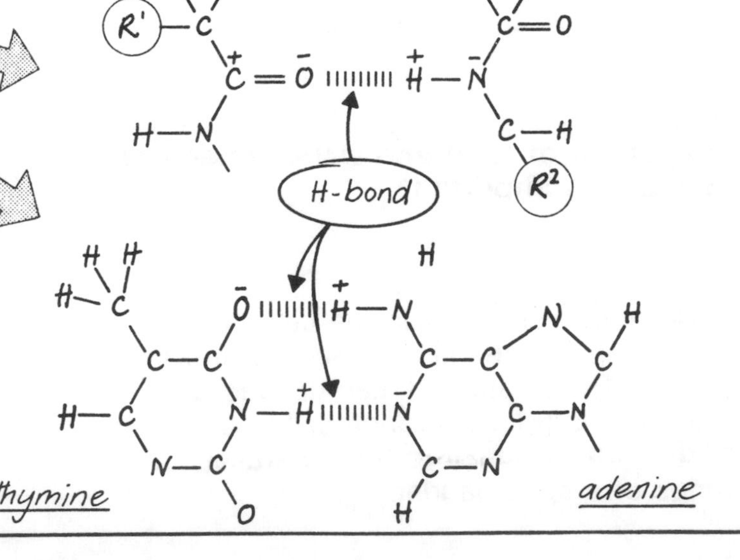

H-bond
thymine
adenine

Although H-bonds are individually very weak, many together will bind two molecules (or two parts of the same molecule) quite strongly. H-bonding holds many proteins in the form of a helix and ties cellulose molecules together to make the tough walls of plant cells.

- In nucleic acids H-bonding will only occur if the two chains are able to fit together closely – this happens only if the A on one chain corresponds to the T on the other, and similarly the G with the C.

This arrangement always occurs in DNA molecules.

- A and T are said to be **complementary** (just as a lock and key are complementary); so are G and C. **It follows that the 2 chains of a DNA double helix are complementary.**

- Because the 2 chains of a **DNA** double helix are complementary, the number of A molecules in the chains must equal the number of T molecules, and the number of G molecules must equal the number of C molecules.

This has been found true of all DNA double helices so far studied.

- **RNA** – occurs both in the nucleus, but more significantly in the cytoplasm of the cell. Complementary chains are *not* found as in the case of DNA, but parts of a strand may form a double helix. Where this occurs the parts of the molecule which 'spiral' together have complementary base pairs.

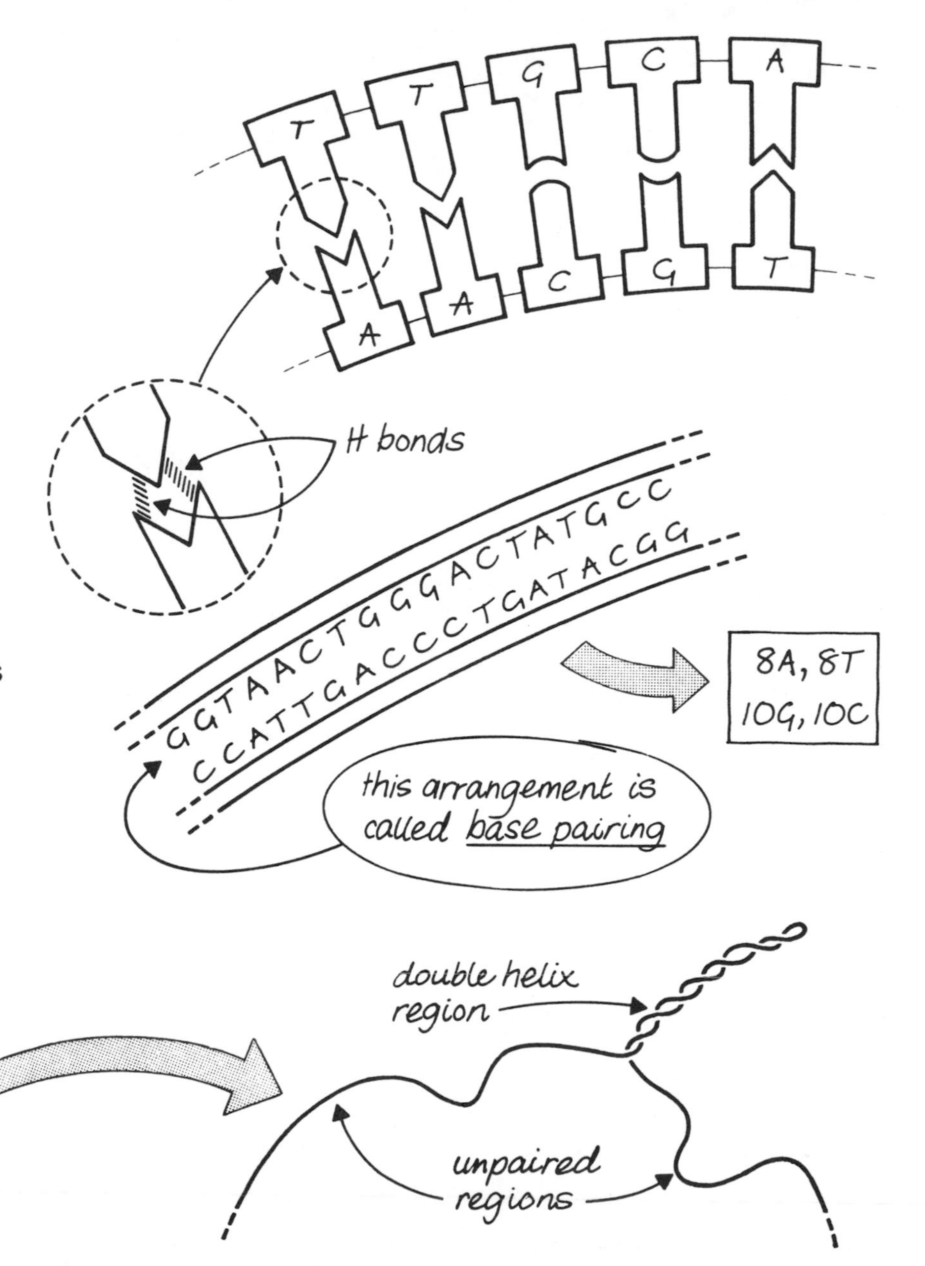

T
T
G
C
A
A
A
C
G
T
H bonds
GGTAACTGGGACTATGCC
CCATTGACCCTGATACGG
8A, 8T
10G, 10C
this arrangement is called base pairing
double helix region
unpaired regions

Part 2 The Catalysts of Metabolism

Enzymes cause biochemical processes in living tissues to happen

- Enzymes are **catalysts**, so act at very low concentration and increase the rate of a chemical reaction.
- Enzymes convert **substrates** into **products**.
- Enzymes act according to the fundamental laws of chemistry.
- All enzymes are proteins, so are 'destroyed' (denatured) by extreme heat (e.g. boiling), by strong acids, or by strong alkalies.
- Any one enzyme will act on only one, or a few chemically related, compounds – it is said to be **specific** for its substrate(s) [≡ it shows **specificity**].
- An enzyme molecule is often very much larger than a molecule of its substrate: enzymes have molecular weights ranging from 10s of 1000s to millions; substrates are usually in the 100s.
- An enzyme molecule has an **active group** which **attacks** the substrate molecule after forming an **enzyme–substrate complex**.

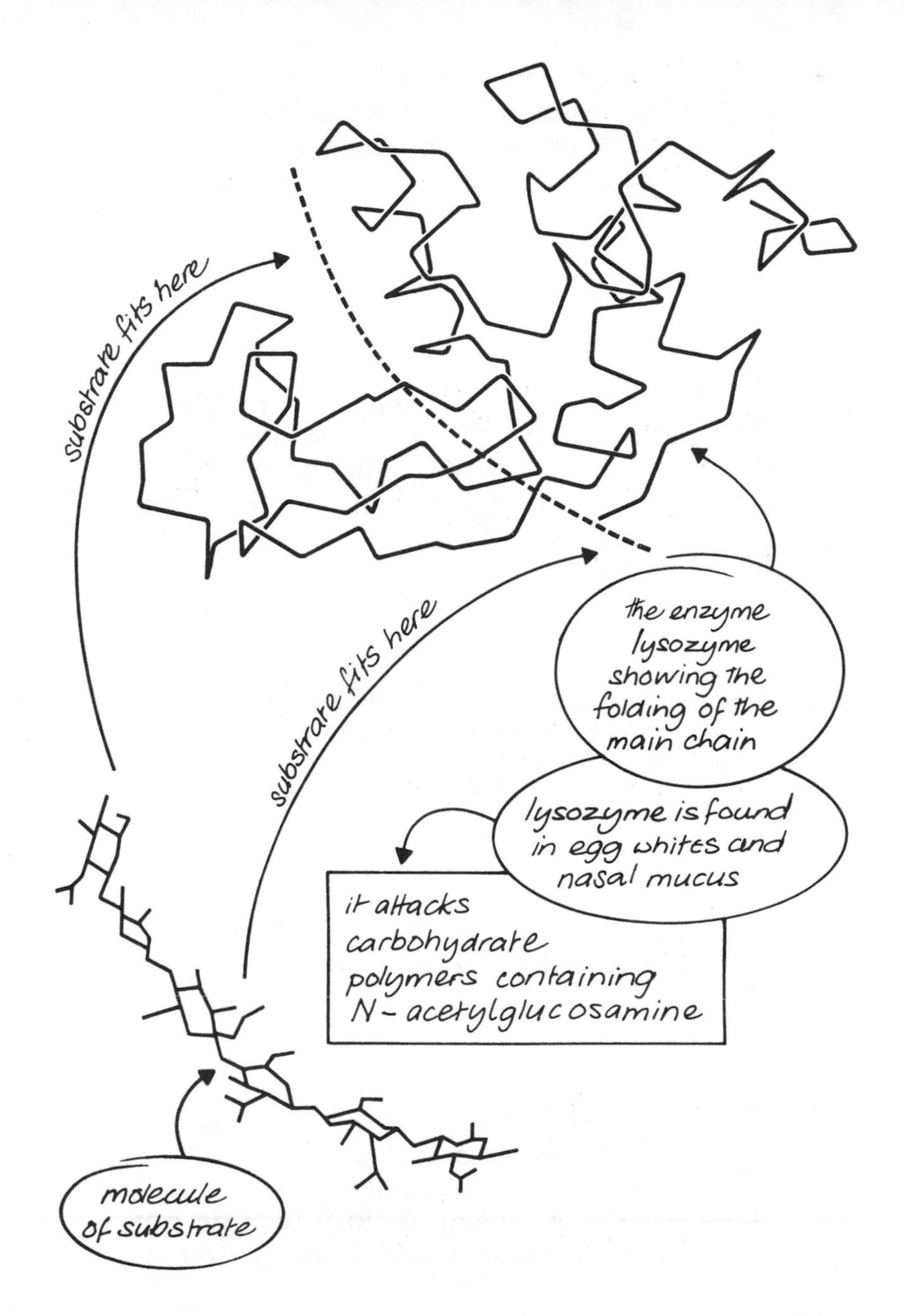

substrate fits here
substrate fits here
the enzyme lysozyme showing the folding of the main chain
lysozyme is found in egg whites and nasal mucus
it attacks carbohydrate polymers containing N-acetylglucosamine
molecule of substrate

- Analysis of the shape of a few enzyme molecules has shown that the substrate molecules **fit** on to the enzyme at one site – the **active site.**

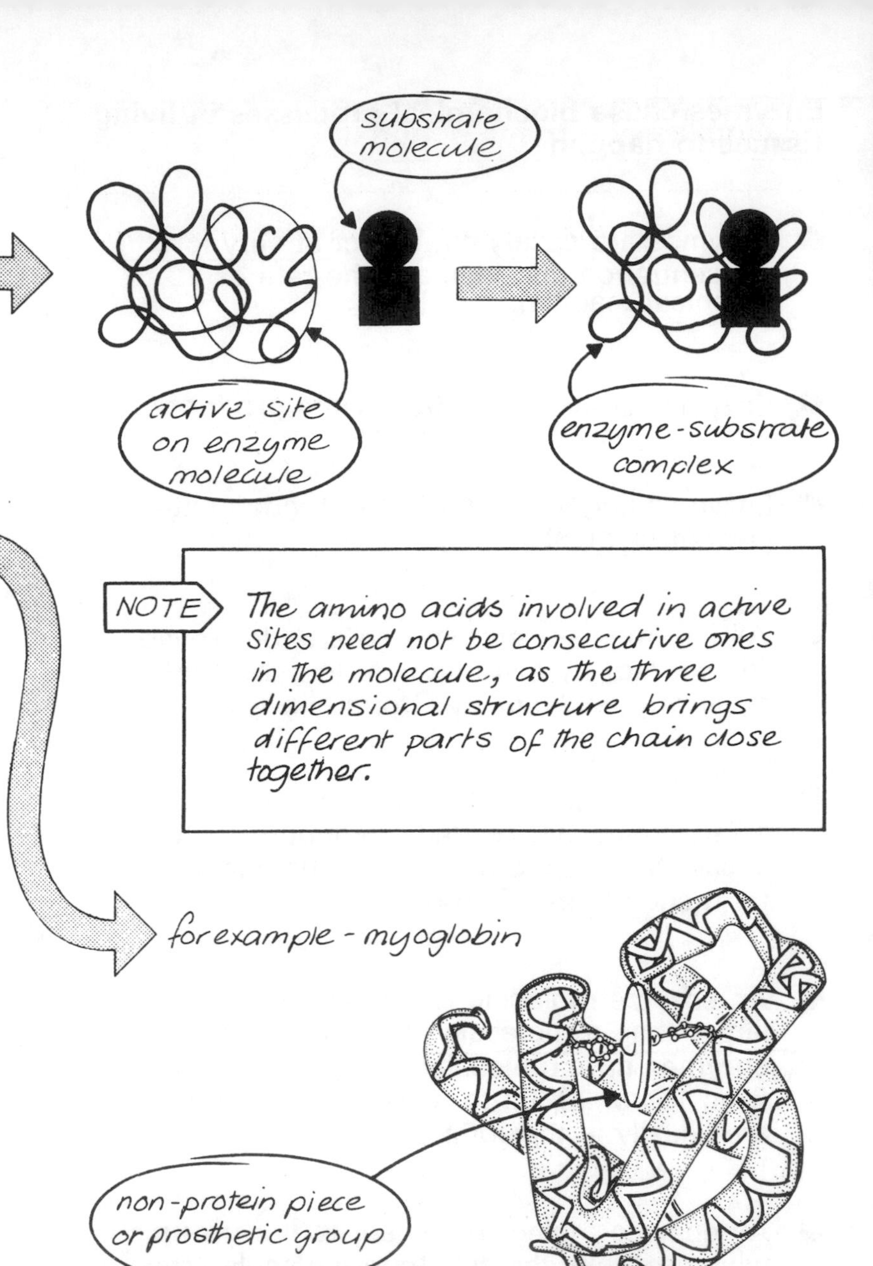

substrate molecule
active site on enzyme molecule
enzyme-substrate complex
NOTE
The amino acids involved in active sites need not be consecutive ones in the molecule, as the three dimensional structure brings different parts of the chain close together.
for example - myoglobin
non-protein piece or prosthetic group

- Some enzymes (not all) need a **non-protein** 'assistant' compound in order to act. If this compound is permanently bound to the enzyme molecule, the enzyme is a **conjugated protein**, and the non-protein piece is called the **prosthetic group**.

- If the non-protein compound is not firmly bound to the protein molecule, but merely needs to be present in the enzyme/substrate mixture, it is called a **co-enzyme**.

- Thousands of different types of **enzyme activity** are known to exist in the living world.

- **Most** enzymes have (long) names which end in -**ase** (invertase, triose phosphate dehydrogenase). A **few** end in -**in** (pepsin, chymotrypsin). A **very few** do not conform to either of these rules (lysozyme).

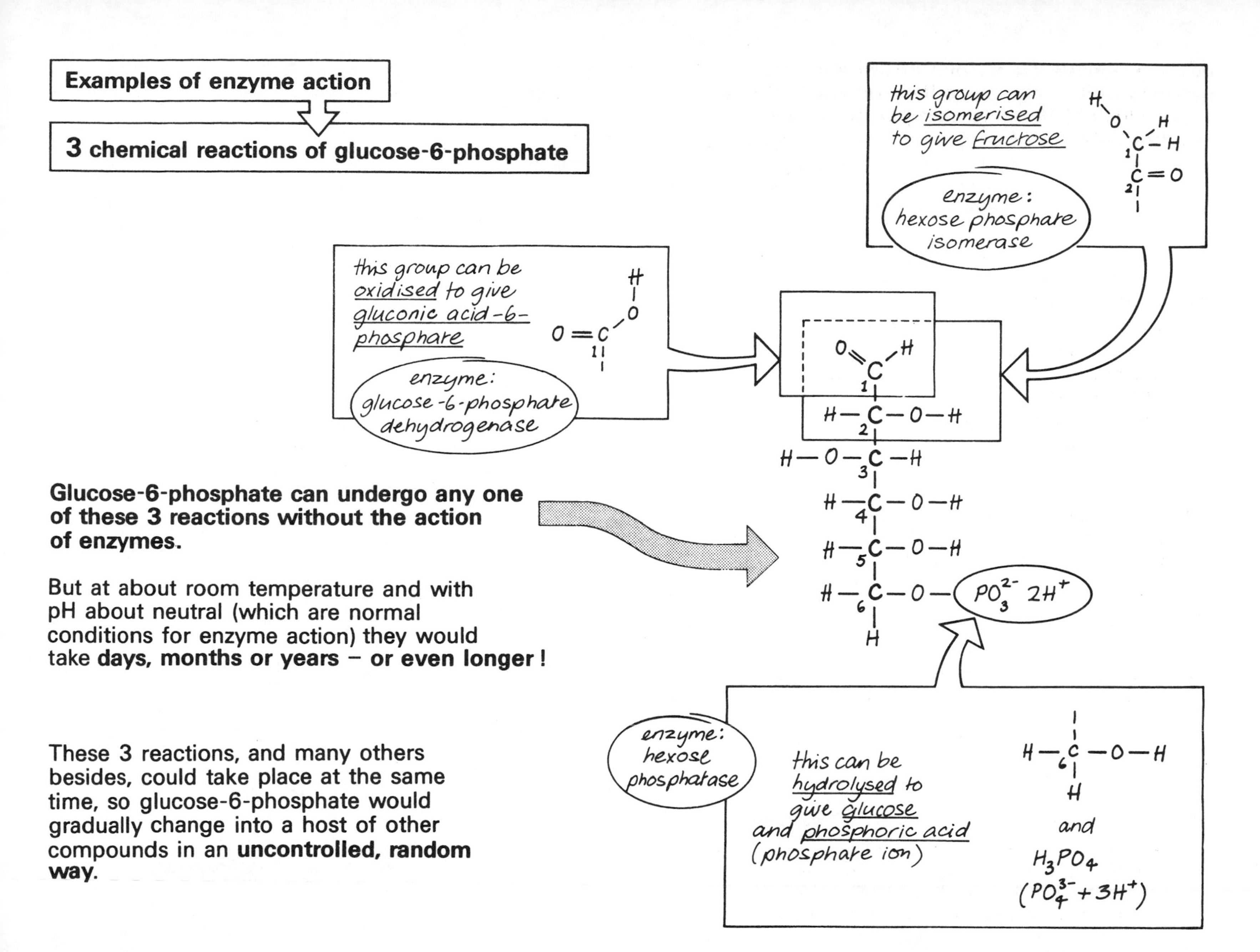

Examples of enzyme action
3 chemical reactions of glucose-6-phosphate
this group can be isomerised to give fructose
enzyme: hexose phosphate isomerase
this group can be oxidised to give gluconic acid-6-phosphate
enzyme: glucose-6-phosphate dehydrogenase
Glucose-6-phosphate can undergo any one of these 3 reactions without the action of enzymes.
But at about room temperature and with pH about neutral (which are normal conditions for enzyme action) they would take days, months or years – or even longer!
These 3 reactions, and many others besides, could take place at the same time, so glucose-6-phosphate would gradually change into a host of other compounds in an uncontrolled, random way.
enzyme: hexose phosphatase
this can be hydrolysed to give glucose and phosphoric acid (phosphate ion)
and
H_3PO_4
$(PO_4^{3-} + 3H^+)$
PO_3^{2-} $2H^+$

In the presence of any one of the enzymes, the corresponding reaction would take place very rapidly (minutes, seconds or even shorter times!) demonstrating both the **specificity** and the **catalytic nature** of enzyme action. The final equilibrium of the substrate and product of any one reaction would be the same as if that reaction had taken place by **any other chemical means (enzymes do not change the laws of chemistry)**.

For instance: consider the isomerisation of glucose-6-phosphate to fructose-6-phosphate:

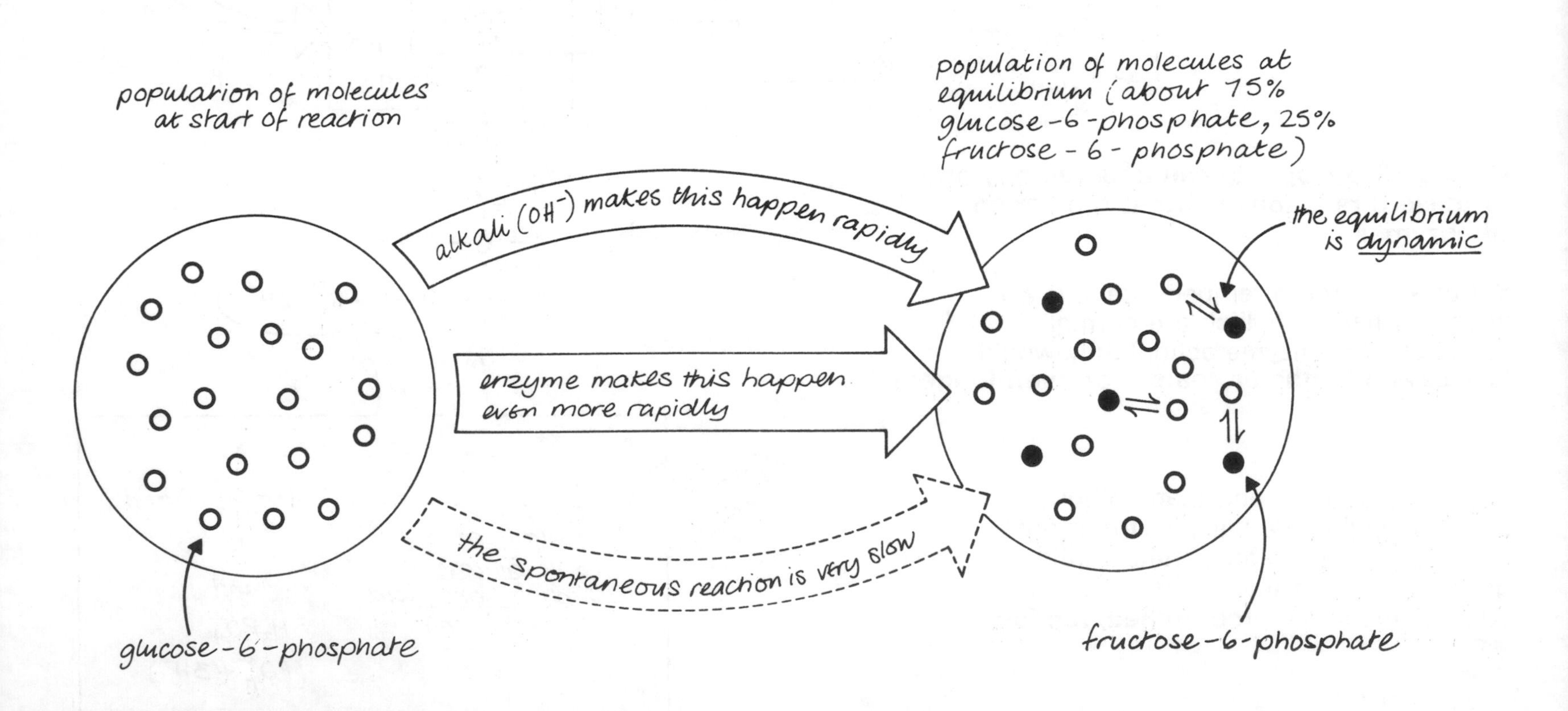

population of molecules at start of reaction
population of molecules at equilibrium (about 75% glucose-6-phosphate, 25% fructose-6-phosphate)
alkali (OH^-) makes this happen rapidly
enzyme makes this happen even more rapidly
the spontaneous reaction is very slow
the equilibrium is dynamic
glucose-6-phosphate
fructose-6-phosphate

More about enzymes

Enzymes are affected by the conditions they find themselves in, such as:

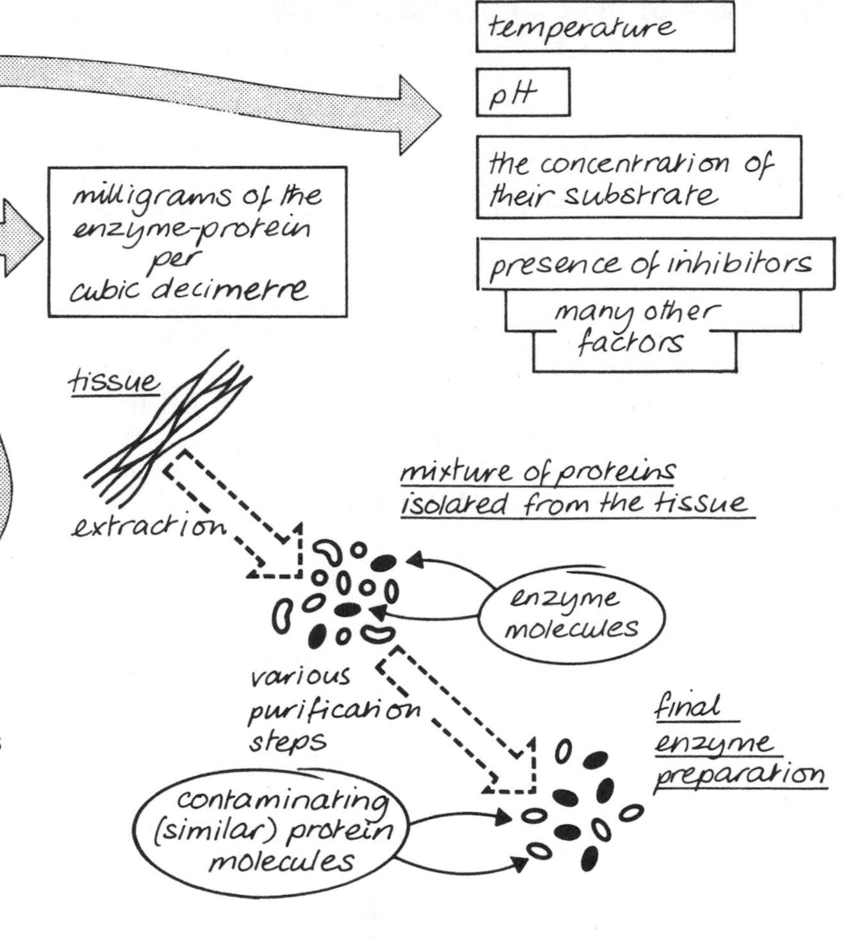

temperature
pH
the concentration of their substrate
presence of inhibitors
many other factors
milligrams of the enzyme-protein per cubic decimetre
tissue
extraction
mixture of proteins isolated from the tissue
enzyme molecules
various purification steps
final enzyme preparation
contaminating (similar) protein molecules

We can find out how an enzyme reacts to these factors by measuring the changes in enzyme activity when the factors change. BUT – enzyme activity is difficult to measure in living tissue as many other enzymes and other compounds interfere. Enzymes are **extracted** from tissues and **purified**, so that the activity of the **isolated** enzyme can be measured under controlled conditions.
As with any other compound, the **amount** of an enzyme can be measured in moles. But we do not know the exact molecular weight of most enzymes, so they are usually measured in weight.

Often the enzyme-protein is not pure enzyme as it is difficult to separate an enzyme from other proteins which may be chemically similar. This does not matter in most cases as the **contaminating** protein does not interfere with the enzyme reaction.

How much enzyme do we have?

We can measure 'amount' of enzyme by its activity

The **amount of activity** is expressed as the amount of substrate converted to product by a known amount of enzyme preparation in a known period of time.

The amount of enzyme (activity) is often expressed in International Units (I.U.) 'One unit of any enzyme is that amount which will catalyse the transformation of one micromole of the substrate per minute under defined conditions'

The effect of pH on enzyme activity

Extremes of pH irreversibly denature enzyme protein.

Less extreme changes of pH affect enzyme activity reversibily (temporarily). Every enzyme has a pH at which it shows its maximum activity. This pH is known as the **pH optimum** for the enzyme. We usually measure enzyme activity in a buffer which has a pH equal to the pH optimum for the enzyme.

Some typical pH curves

Note 1: The purity of the enzyme preparation usually has no effect upon the pH optimum

Note 2: When expressing activity in I.U.s, the measurements must be made at the optimum pH for obvious reasons. (For instance, if they were made at pH 3 or at pH 11 in any of the cases above, the values in I.U. would all be near zero!)

The effect of temperature on enzyme activity

Some enzymes can withstand quite high temperatures (even up to 100°C) for several minutes – or even longer. But most enzymes are denatured at high temperature – many at temperatures as low as 50–60°C.

The **time of exposure** is important – several hours at fairly low temperatures (even normal room temperature) may be more destructive than a few seconds at temperatures near 100°C – – enzymes extracted from living tissues are therefore usually handled and stored at low temperature (refrigerator, cold room, deep freeze) to prolong their active life.

The rate of enzyme action is affected by change in temperature. Chemical reactions are faster the higher the temperature: **the rate is about doubled for each 10°C rise** and this is true of enzyme reactions also.

This graph shows that the rate of increase of enzyme-catalysed reactions at higher temperatures may be more than offset by the destruction of the enzyme by heat.

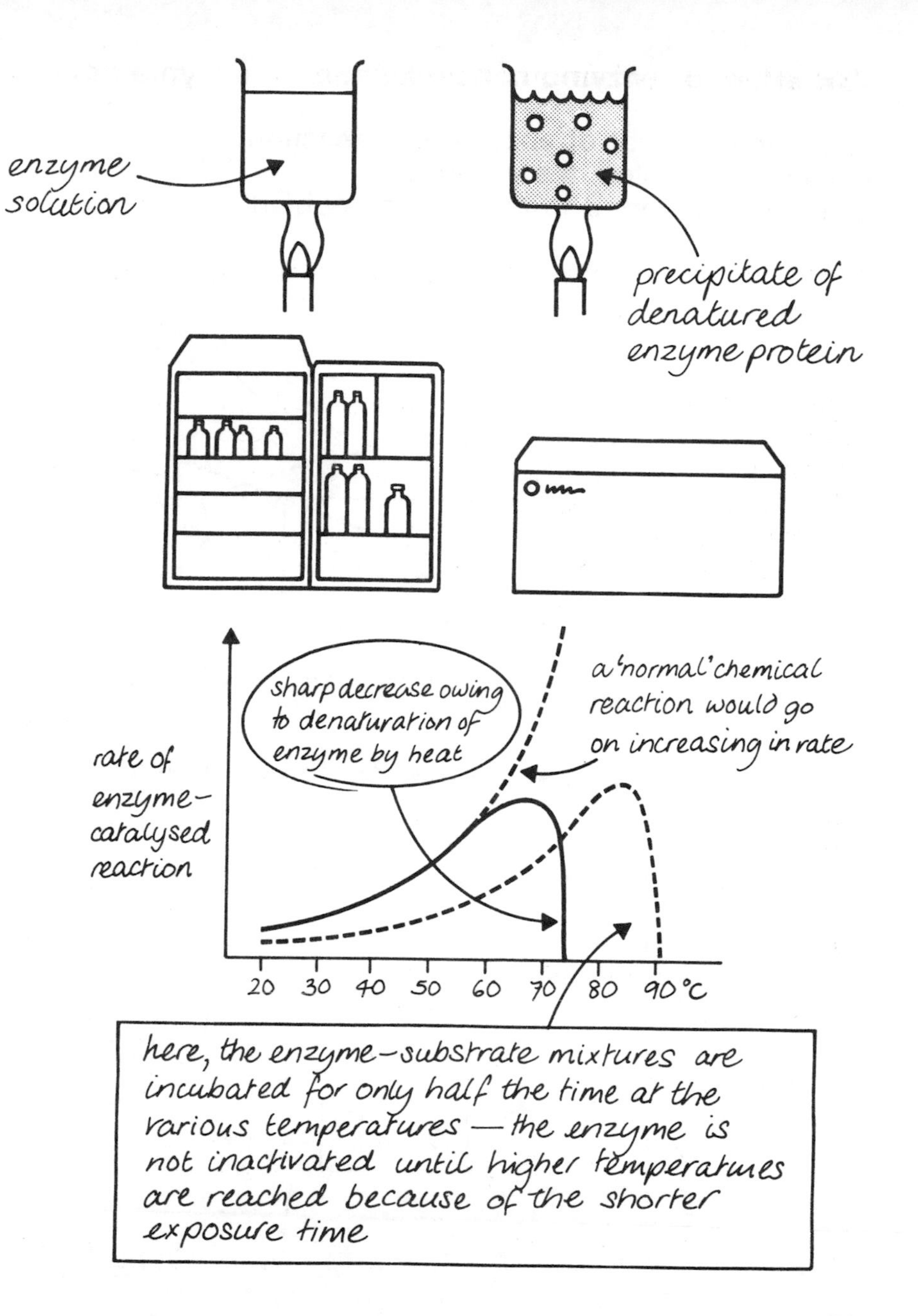

enzyme solution
precipitate of denatured enzyme protein
rate of enzyme-catalysed reaction
sharp decrease owing to denaturation of enzyme by heat
a 'normal' chemical reaction would go on increasing in rate
20 30 40 50 60 70 80 90 °C
here, the enzyme-substrate mixtures are incubated for only half the time at the various temperatures — the enzyme is not inactivated until higher temperatures are reached because of the shorter exposure time

Note > When expressing activity in International Units, the temperature at which measurements are made is 25°C (by international agreement).

The effect of enzyme concentration on enzyme activity

When the amount of enzyme in a reaction mixture is doubled, the amount of substrate converted to product is doubled.

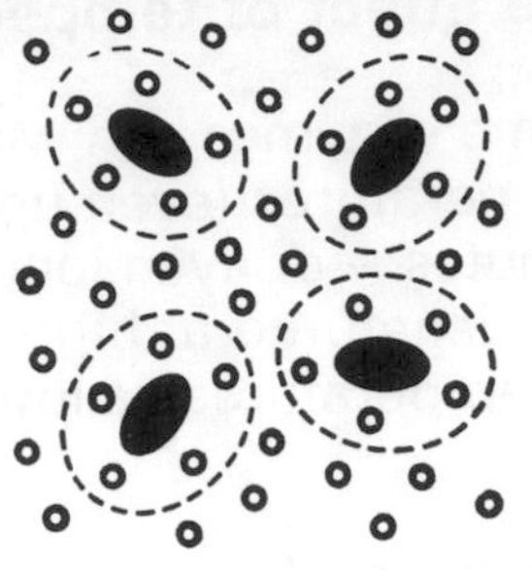

2 enzyme molecules 'capture' (convert) 2 x 5 substrate molecules in a given time

4 enzyme molecules (in the same volume) convert 4 x 5 substrate molecules in the same time

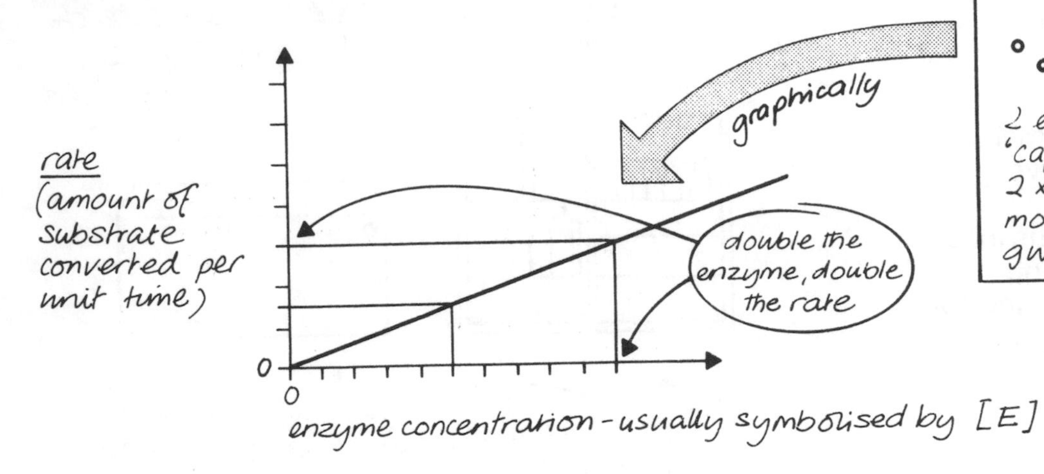

graphically
rate
(amount of substrate converted per unit time)
double the enzyme, double the rate
0
0
enzyme concentration - usually symbolised by [E]

The effect of substrate concentration on enzyme activity

The graph relating substrate concentration to enzyme activity is:

Note the enzyme concentration is kept constant.

plateau

rate (usually symbolised by v)

substrate concentration (usually symbolised by [S])

here the rate is proportional to [S], e.g. doubling the substrate concentration doubles the rate

here the rate is unaffected by [S], e.g. doubling the substrate concentration has no effect upon the rate

Why?

Assume a molecule of enzyme **has the capacity** to convert 100 molecules of substrate per second (at 25°C, at optimum pH). **If** it is 'presented' with only 1 molecule of substrate per second, the rate of reaction would be 1 molecule of substrate converted per second. This happens when the concentration of substrate is **very low** – the substrate molecules are few and far between.

So, if, in our example we double the concentration of substrate so that 2 molecules of substrate present themselves to the enzyme molecule per second, the rate would be 2 molecules of substrate converted per second –

and if we double the concentration again, the rate would increase to 4 molecules of substrate converted per second and if we increase the concentration of substrate to 50 times its original concentration, the rate would increase to 50 molecules of substrate converted per second.

–**So far, the rate is proportional to the concentration of substrate.**

But, when we have increased the substrate concentration so much that the enzyme is now presented with 100 molecules per second, it reaches its **absolute maximum rate**. Doubling the concentration now, so that 200 substrate molecules are presented per second, does not increase the rate as the enzyme molecule cannot cope with more than 100 per second. The enzyme is said to be **saturated** with substrate.

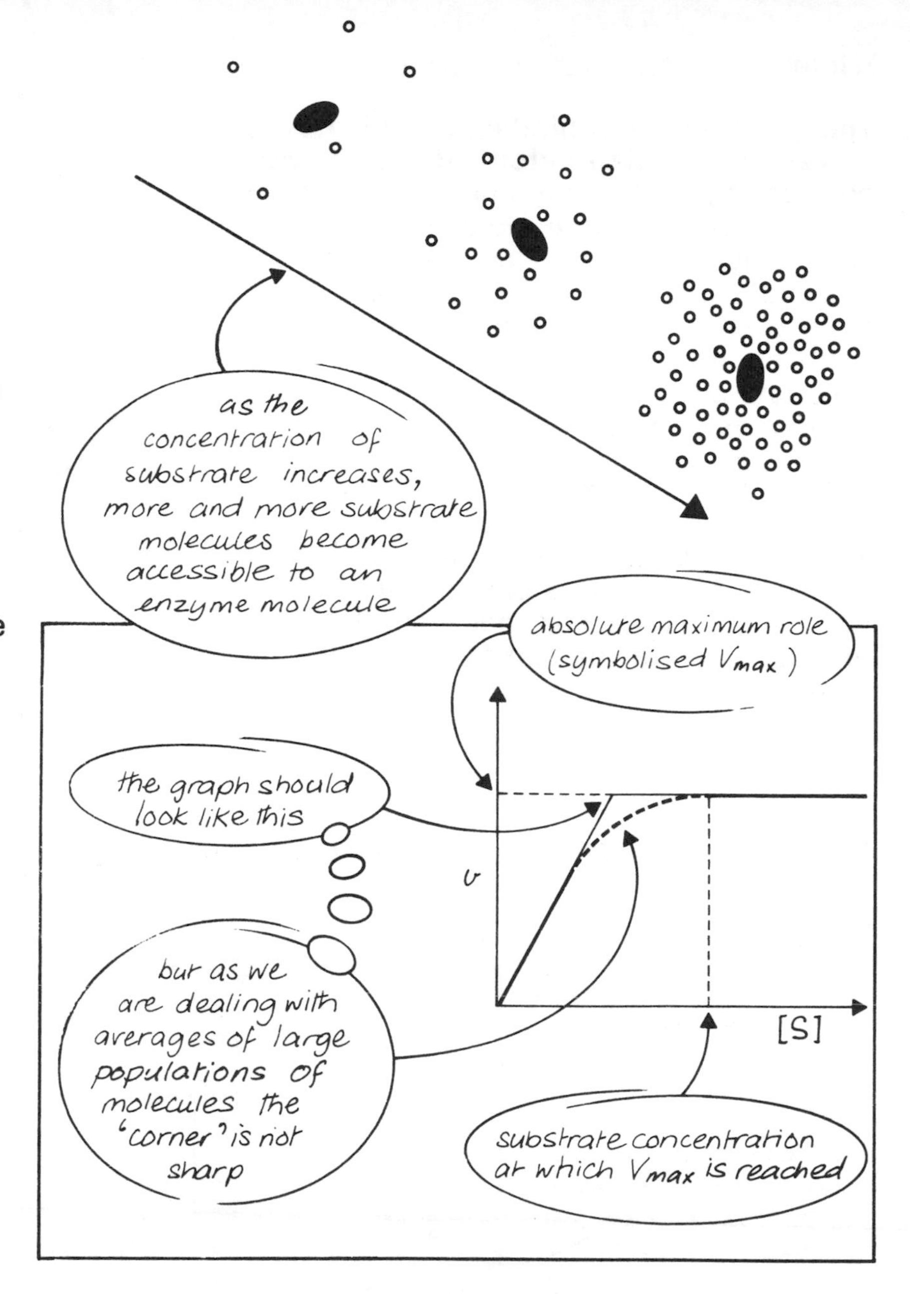
as the concentration of substrate increases, more and more substrate molecules become accessible to an enzyme molecule
absolute maximum role (symbolised V_{max})
the graph should look like this
v
but as we are dealing with averages of large populations of molecules the 'corner' is not sharp
[S]
substrate concentration at which V_{max} is reached

Note

The substrate concentration at which V_{max} is reached is **independent of the enzyme concentration**. For any one enzyme, this is **because** each enzyme molecule in solution acts independently of every other enzyme molecule, so 1 or 2 or 4 or 50 enzyme molecules per unit volume of solution will **each** be affected in the same way by the substrate concentration (or the pH, or the temperature, or any other factor).

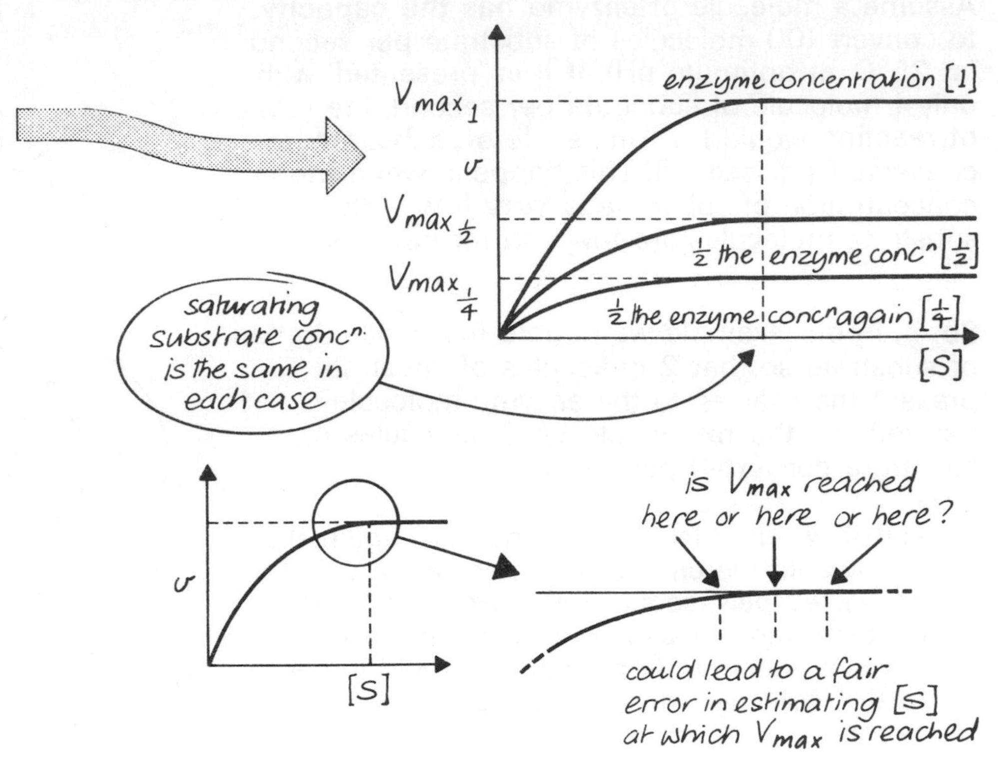

V_{max_1}
enzyme concentration [1]
v
$V_{max_{\frac{1}{2}}}$
½ the enzyme concⁿ [½]
$V_{max_{\frac{1}{4}}}$
½ the enzyme concⁿ again [¼]
[S]
saturating substrate concⁿ is the same in each case
v
[S]
is V_{max} reached here or here or here?
could lead to a fair error in estimating [S] at which V_{max} is reached

A practical consideration. It is difficult to measure the substrate concentration at which V_{max} is reached, because of the gradual slope of the curve reaching the plateau.

So for convenience ½ V_{max} is measured **because**

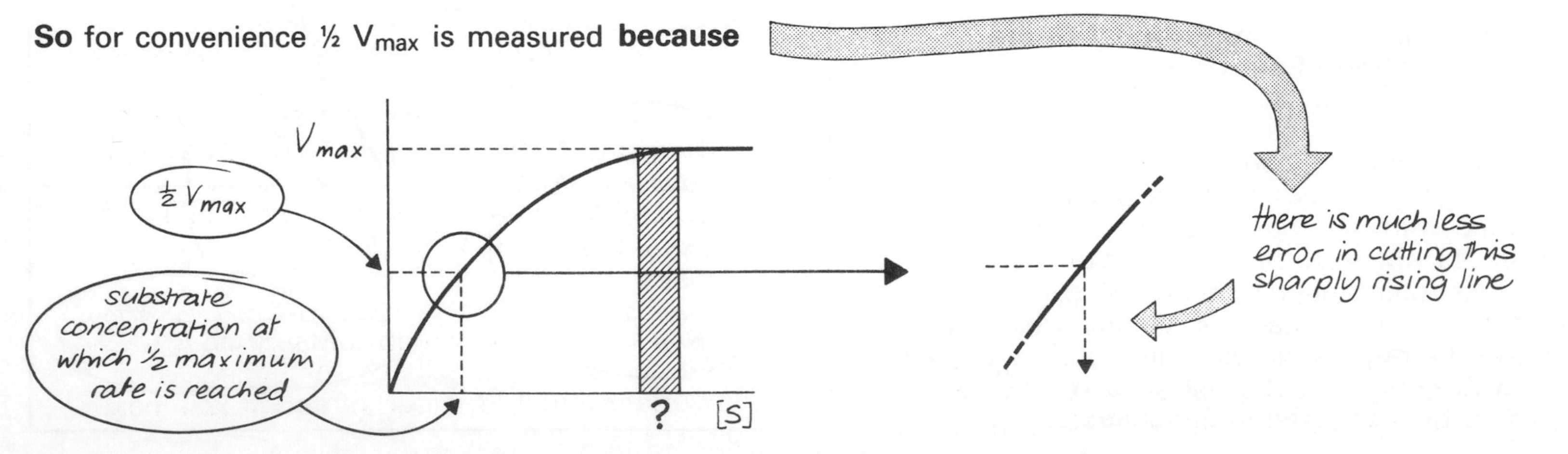

V_{max}
½ V_{max}
substrate concentration at which ½ maximum rate is reached
?
[S]
there is much less error in cutting this sharply rising line

The value of [S] at which ½ V_{max} is reached is called the **Michaelis Constant** and its symbol is K_m.

Note 1 > the graph shows that the K_m is much less than ½ the [S] at which V_{max} is reached.

Note 2 > the K_m is constant for an enzyme regardless of its concentration.

Note 3 > if an enzyme has 2 (or more) substrates, e.g. A+B ⟶ product, it has 2 (or more) K_m's – one for each substrate.

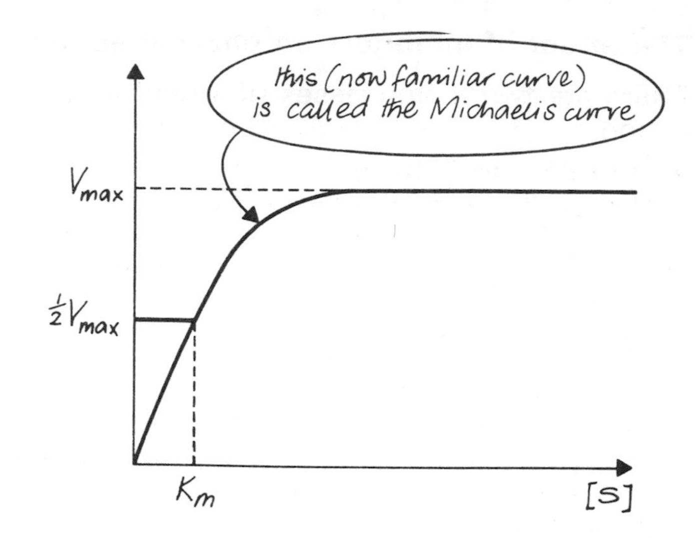
this (now familiar curve) is called the Michaelis curve
V_{max}
$\frac{1}{2}V_{max}$
K_m
[S]

What use is the Michaelis Constant ?

1. It is a characteristic of an enzyme (e.g. like the boiling point of a liquid) – but note that more than one enzyme may have the same K_m.

2. It tells us about the **affinity** of the enzyme for the substrate. An enzyme which has a **low** K_m has a **high** affinity for its substrate, because it becomes **saturated** (works at maximum rate) at a **low substrate concentration**.

3. Enzymes with **low** K_m (high affinity) are **highly important in metabolism**: low K_m values are of the order of $\frac{1}{1{,}000{,}000}$ Molar $= 10^{-6}$ M, or even lower (10^{-7} M, or 10^{-8} M). Enzymes with high K_m (lower affinity) are **less important in metabolism**: these K_m values are of the order of $\frac{1}{100}$ Molar $= 10^{-2}$ M, or even higher (10^{-1} M). This is only a rough guide as there is a continuous spectrum of K_m values, but it is a useful rule of thumb nevertheless for people investigating enzymes.

4. When expressing enzyme activity in I.U.'s we have already seen that the measurements are made at optimum pH and at a standard temperature (25°C). **They must also be made at saturating substrate concentration** for obvious reasons: the same enzyme could give very different values if the measurements were made on different parts of the Michaelis curve.

5. If an inhibitor is acting on the enzyme, its effect upon the K_m tells us something important about its mode of action (see next page) ⟶

The effect of inhibitors on enzyme action

There are two main classes of inhibitors of enzyme action.

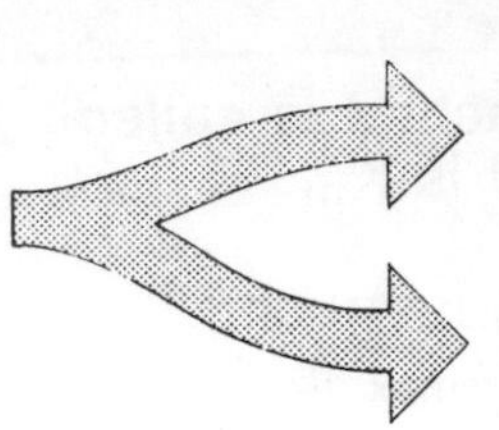

irreversible inhibitors

reversible inhibitors

Irreversible inhibitors

When the enzyme is exposed to the inhibitor, it forms a covalent bond which cannot be broken easily. The enzyme inhibitor complex is permanently inactive: the enzyme has been **poisoned**.

an example

The nerve poison **DFP**

active enzyme + **DFP**
(necessary for nerve action)
↓ *irreversible*
inactive **DFP**-*enzyme complex*

Reversible inhibitors

Here the inhibitor readily dissociates from the enzyme leaving its activity unimpaired. Reversible inhibitors can be sudbivided into 2 groups.

① **Competitive inhibitors compete** with the substrate molecules for the active site on the enzyme molecule, **or** put differently, the enzyme molecule 'recognises' the inhibitor as a substrate but then is unable to convert it to product.

enzyme
inhibitor
enzyme-inhibitor complex
enzyme
unchanged inhibitor

In a mixture of **substrate** and **inhibitor** the enzyme will convert less substrate molecules than it would do otherwise, because for part of the time each enzyme molecule is occupied by inhibitor instead of substrate. Increasing the inhibitor concentration relative to the substrate **increases** the inhibition.

Increasing the substrate relative to the inhibitor makes it more likely that substrate molecules will combine with the enzyme, so decreasing the inhibition: the substrate and inhibitor **compete** for the enzyme.

enzyme + *substrate* → *enzyme-substrate complex* → *enzyme* + *products*

Note 1

Molecules of a competitive inhibitor are often chemically similar to molecules of the substrate.

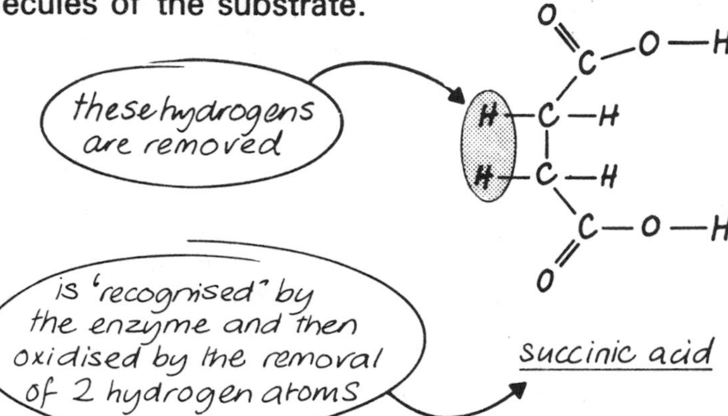

these hydrogens are removed
is 'recognised' by the enzyme and then oxidised by the removal of 2 hydrogen atoms
succinic acid

An example

Succinic acid dehydrogenase is inhibited competitively by malonic acid.

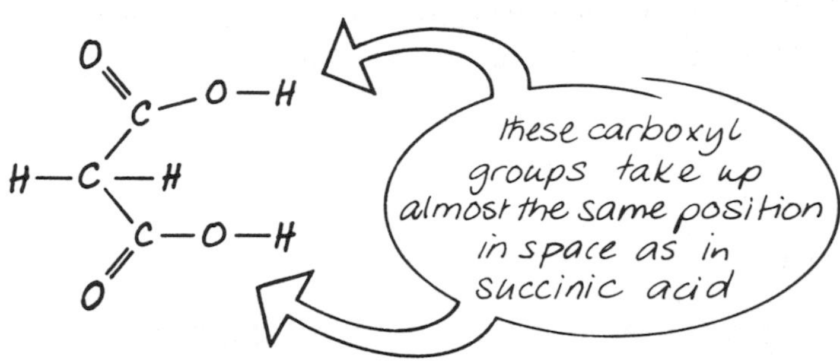

these carboxyl groups take up almost the same position in space as in succinic acid
malonic acid

is 'recognised' by the enzyme, but the active group cannot oxidise it, so it 'falls off' the enzyme unchanged – meanwhile it has prevented a molecule of succinic acid being oxidised

Note 2

The Michaelis curve is changed when a competitive inhibitor is present.

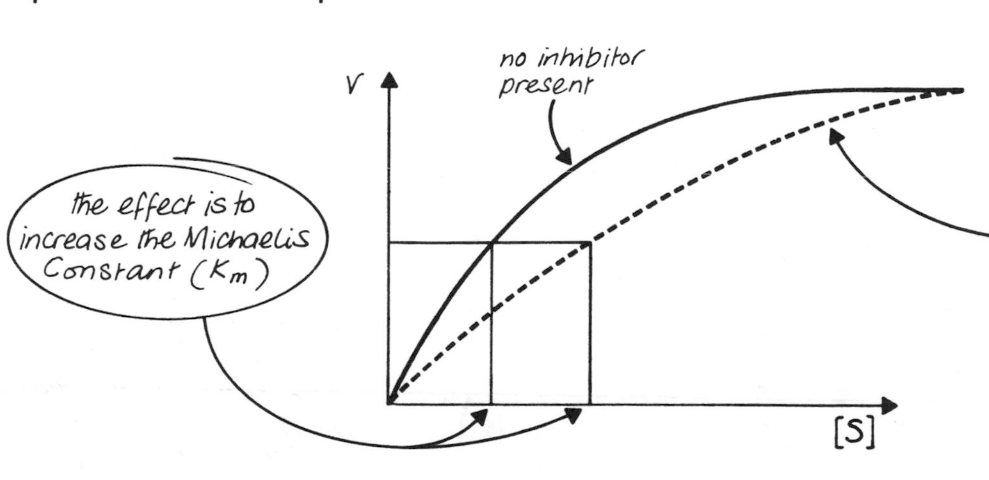

V
no inhibitor present
the effect is to increase the Michaelis Constant (K_m)
[S]

*inhibitor present (at constant concentration). Its effect is overcome by **very high** [S] competing for the enzyme. So V_{max} is the same (but is reached at a higher [S]).*

② Non competitive inhibitors

Many inhibitors of enzymes are known which are reversible but non-competitive. They affect a different site on the enzyme molecule, but in doing so prevent its activity.

substrate fits here

non-competitive inhibitor molecule changes the enzyme molecule so that substrate will not 'fit'

Note 1

Here the amount of inhibition depends upon the concentration of inhibitor regardless of the substrate concentration, as the 2 types of molecule are not 'competing' for the same site on the enzyme molecule.

2 different sites involved

Note 2

The Michaelis curve is also changed, but differently from the previous case.

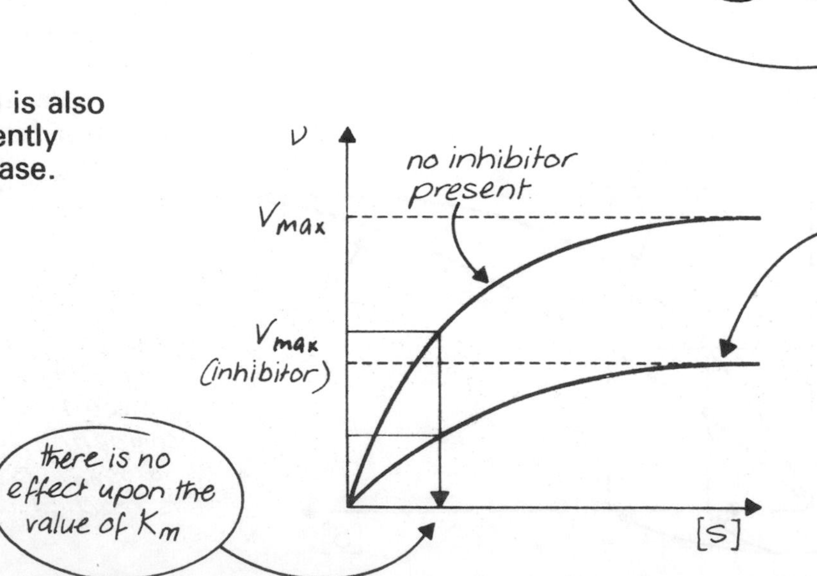

v
no inhibitor present
V_{max}
V_{max} (inhibitor)
there is no effect upon the value of K_m
[S]

inhibitor present at constant concentration. Its effect is not overcome by [S]. Although the inhibited molecules do not convert substrate at all, the uninhibited ones react to [S] in the normal way. So V_{max} is <u>reduced</u> (but is reached at the same [S])

Note 3

This type of inhibition may be used in living tissue to 'turn off' a reaction sequence when the final product is produced more rapidly than it is used. The change of 'shape' of the enzyme to an inactive form is called the **allosteric effect**.

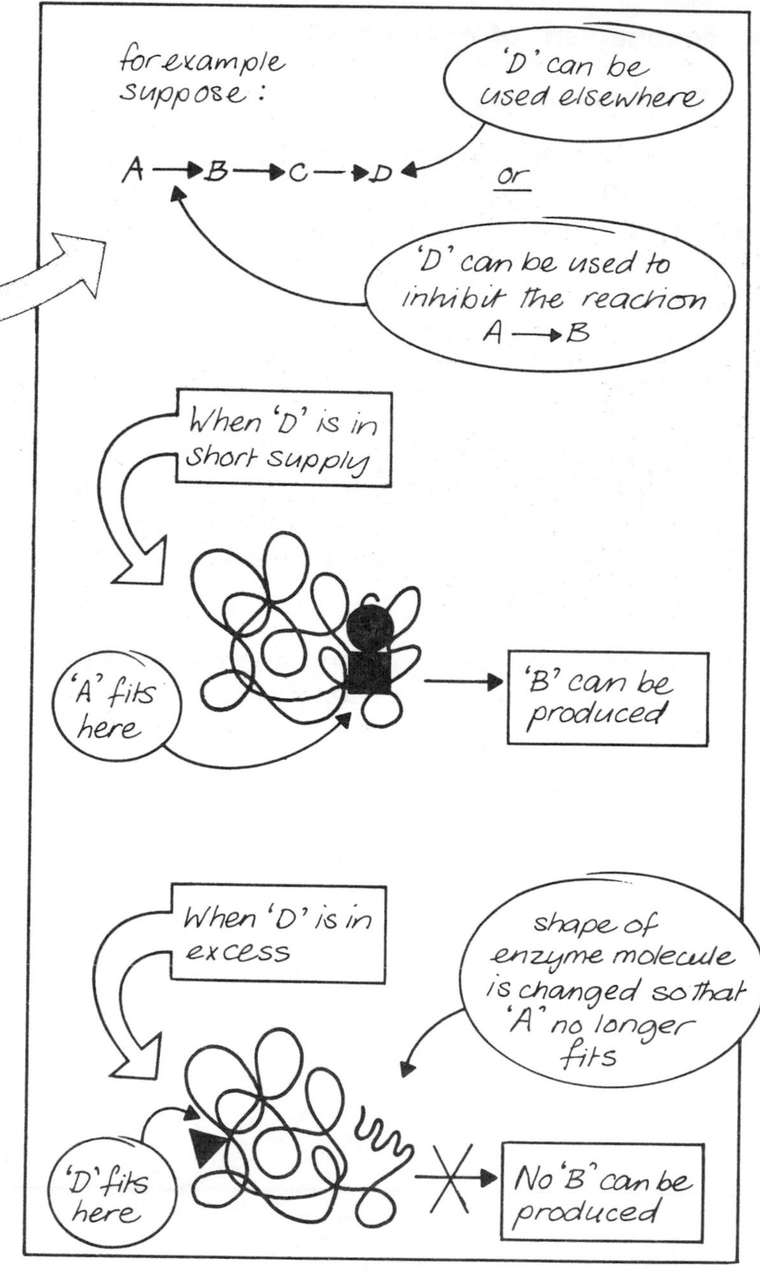

for example suppose :
'D' can be used elsewhere
A → B → C → D
or
'D' can be used to inhibit the reaction A → B
When 'D' is in short supply
'A' fits here
'B' can be produced
When 'D' is in excess
shape of enzyme molecule is changed so that 'A' no longer fits
'D' fits here
No 'B' can be produced

Turnover number

We have seen that the amount of substrate an enzyme converts under particular conditions depends upon certain **intrinsic characters** of the enzyme molecule (its Michaelis constant for instance, and its pH optimum). The **turnover number** is another important intrinsic character. Even when saturated with substrate and at optimum pH, some enzymes will act faster than others – i.e. will convert more molecules of substrate per molecule of enzyme in a given time. So a turnover number can be defined as **moles of substrate converted per mole of enzyme per second** (at optimum pH, 25°C and saturating substrate concentration).

The actual values vary enormously for different enzymes, for instance

Succinic acid dehydrogenase has a turnover number of

1,150

Carbonic anhydrase has a turnover number of

36,000,000

Note The turnover number is **independent of the Michaelis constant**: an enzyme with a high turnover could have either a high or low K_m.

Mechanism of enzyme action

Two main questions can be asked:

How is it possible that a chemical reaction which may take years to occur can be shortened to take only seconds or even fractions of a second?

What happens at molecular level between the enzyme and substrate molecules?

A chemical reaction can only happen spontaneously if there is a decrease in **free energy**. Free energy may be looked upon as energy **capable of doing work**.

So A+B ⟶ AB will happen spontaneously only if the free energy content of AB is lower than A+B.

If AB contains more free energy than A + B the reaction must be driven 'uphill' (up to a higher energy level) and the reaction will not happen spontaneously, but will need an input of **heat** or some other form of energy (e.g. **chemical bond energy**). This can be done by a **coupled reaction**.

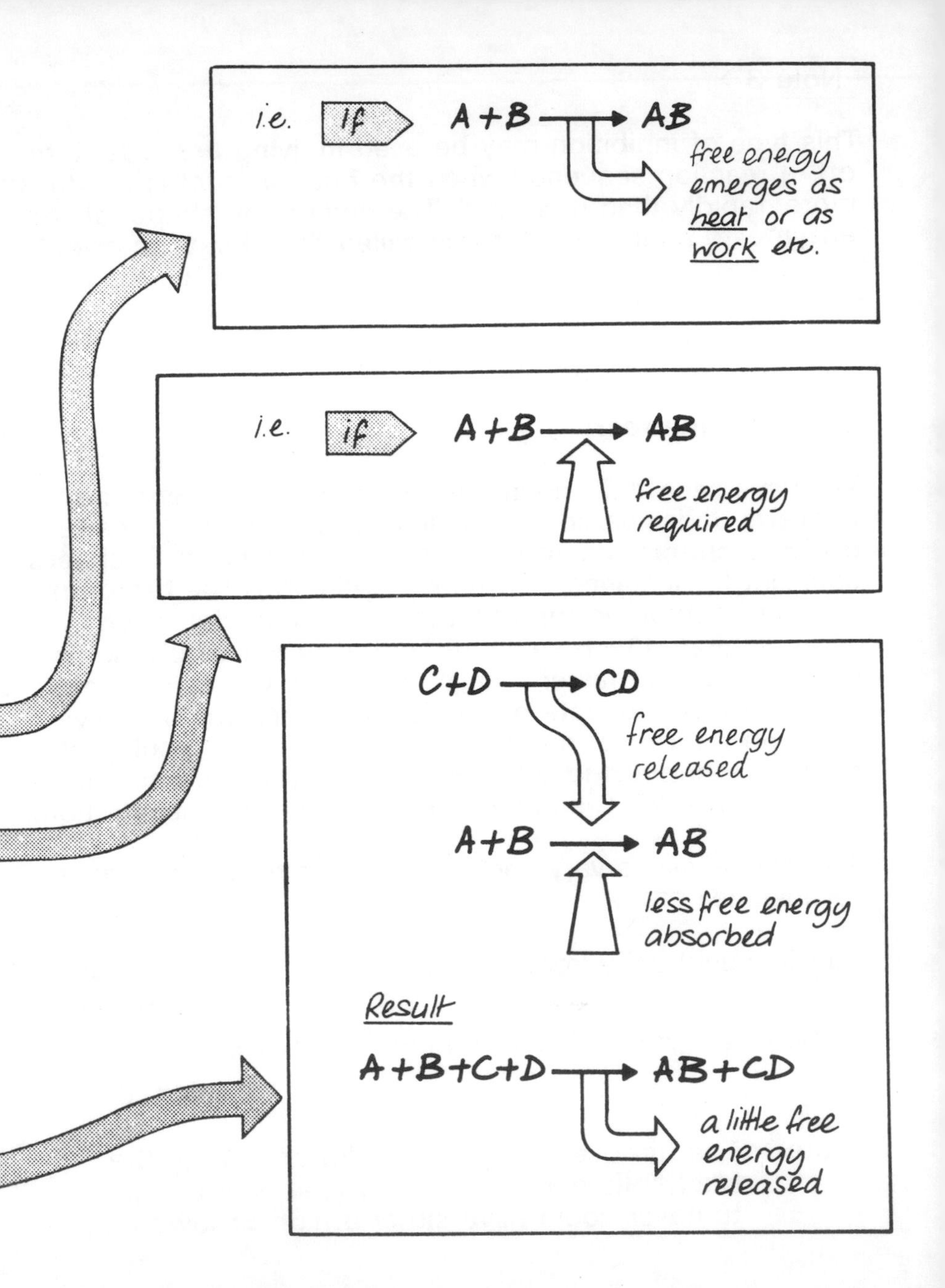

i.e. if A+B ⟶ AB
free energy emerges as heat or as work etc.
i.e. if A+B ⟶ AB
free energy required
C+D ⟶ CD
free energy released
A+B ⟶ AB
less free energy absorbed
Result
A+B+C+D ⟶ AB+CD
a little free energy released

Even if a reaction appears to be spontaneous from the free energy relationships, **it need not occur or it may take place only very slowly** i.e. the **rate** of reaction may be low.

Turning back to the chemical reactions,

A and B may have energy barriers to overcome before they can react A and B must be activated (pushed uphill) to a higher level before they will react. This level is called the **activation energy**.

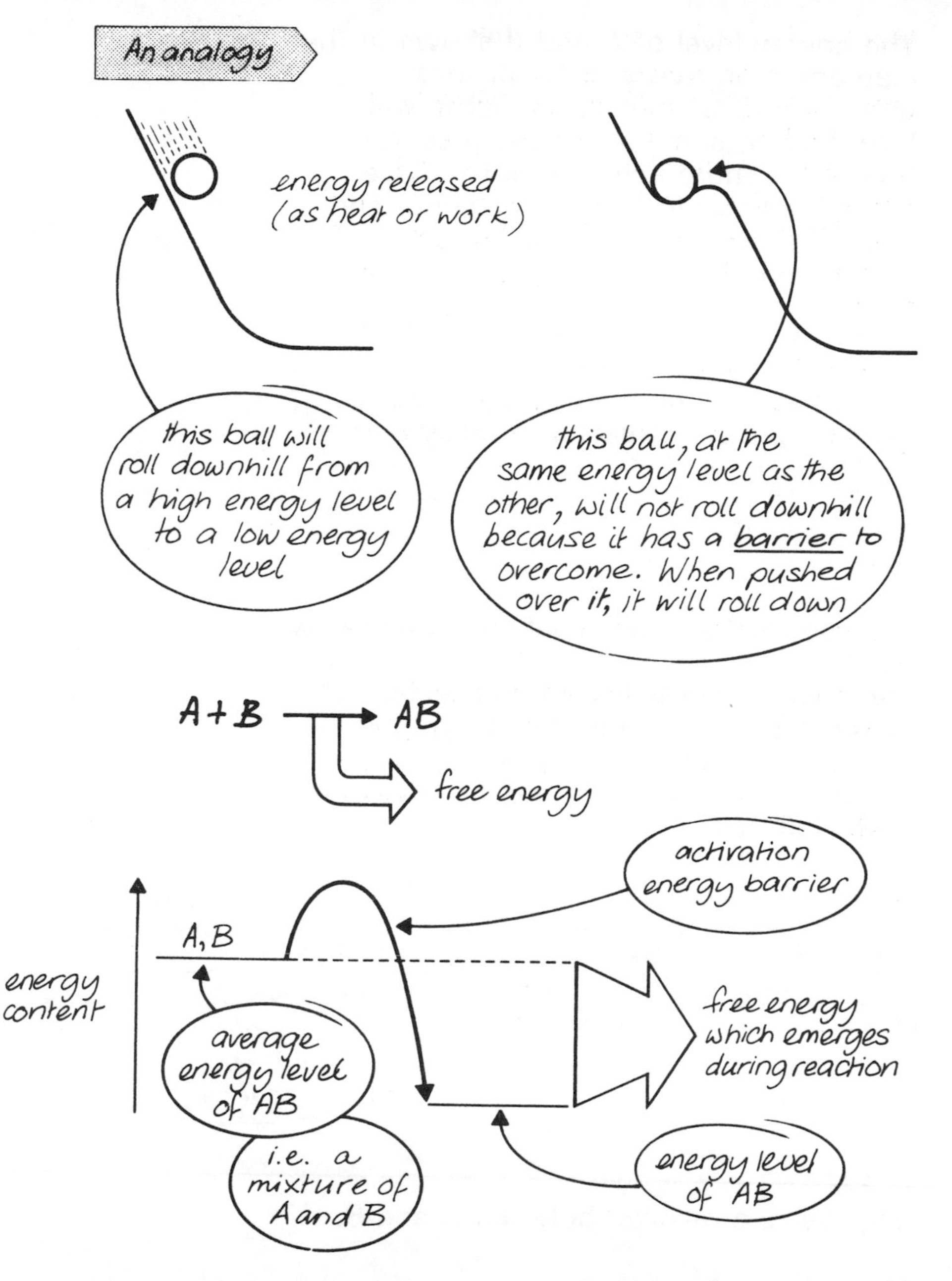

An analogy
energy released (as heat or work)
this ball will roll downhill from a high energy level to a low energy level
this ball, at the same energy level as the other, will not roll downhill because it has a barrier to overcome. When pushed over it, it will roll down
A+B → AB
free energy
activation energy barrier
A, B
energy content
average energy level of AB
i.e. a mixture of A and B
free energy which emerges during reaction
energy level of AB

The energy level of A and B shown in the diagram is an **average** for billions upon billions of molecules. Some will have higher, some lower energies. A few of the more active ones will have enough energy to 'jump' the barrier and react. The **rate** of the reaction will depend on how many do this per unit time.

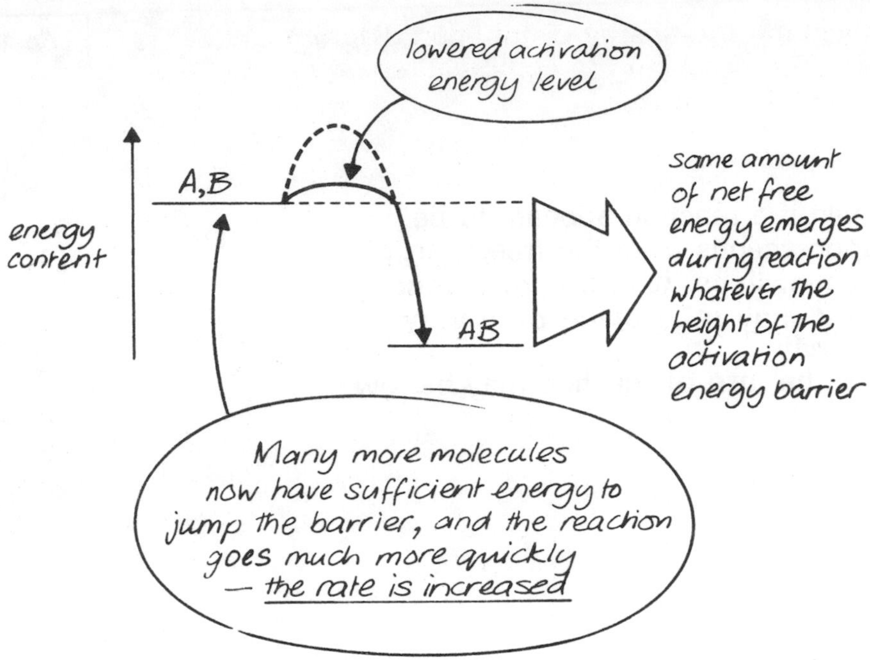

> Enzymes and all other catalysts act by lowering the **activation energy** barrier.

We can illustrate the situation in another way:

The energy distribution of a collection of molecules may be represented graphically. The 'shape' of this energy distribution takes the form of a **'normal'** or **Gaussian distribution**.

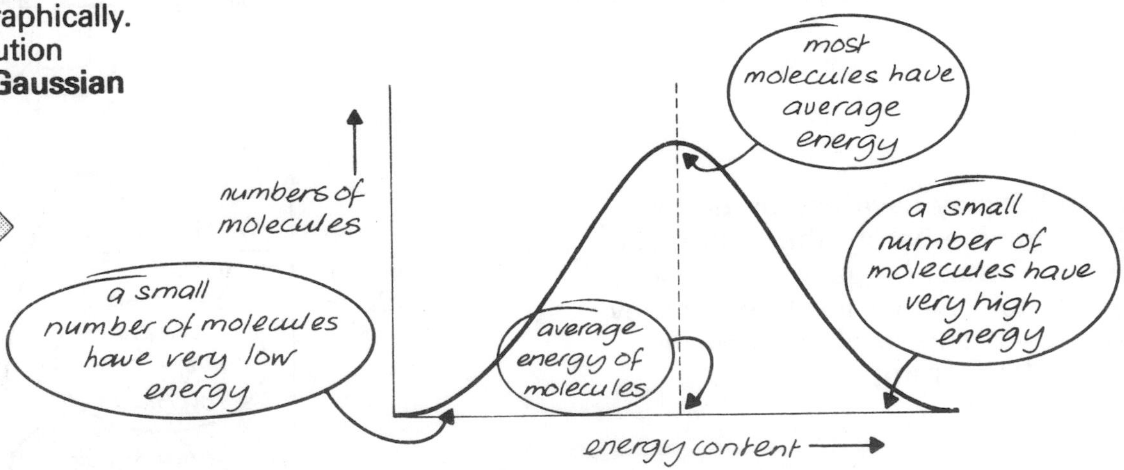

NOW
consider the reaction between A and B.

This time the energy distribution diagram is concerned with the combined energy levels of A and B.

In the presence of an enzyme (or other catalyst) the energy distribution diagram now looks like this:

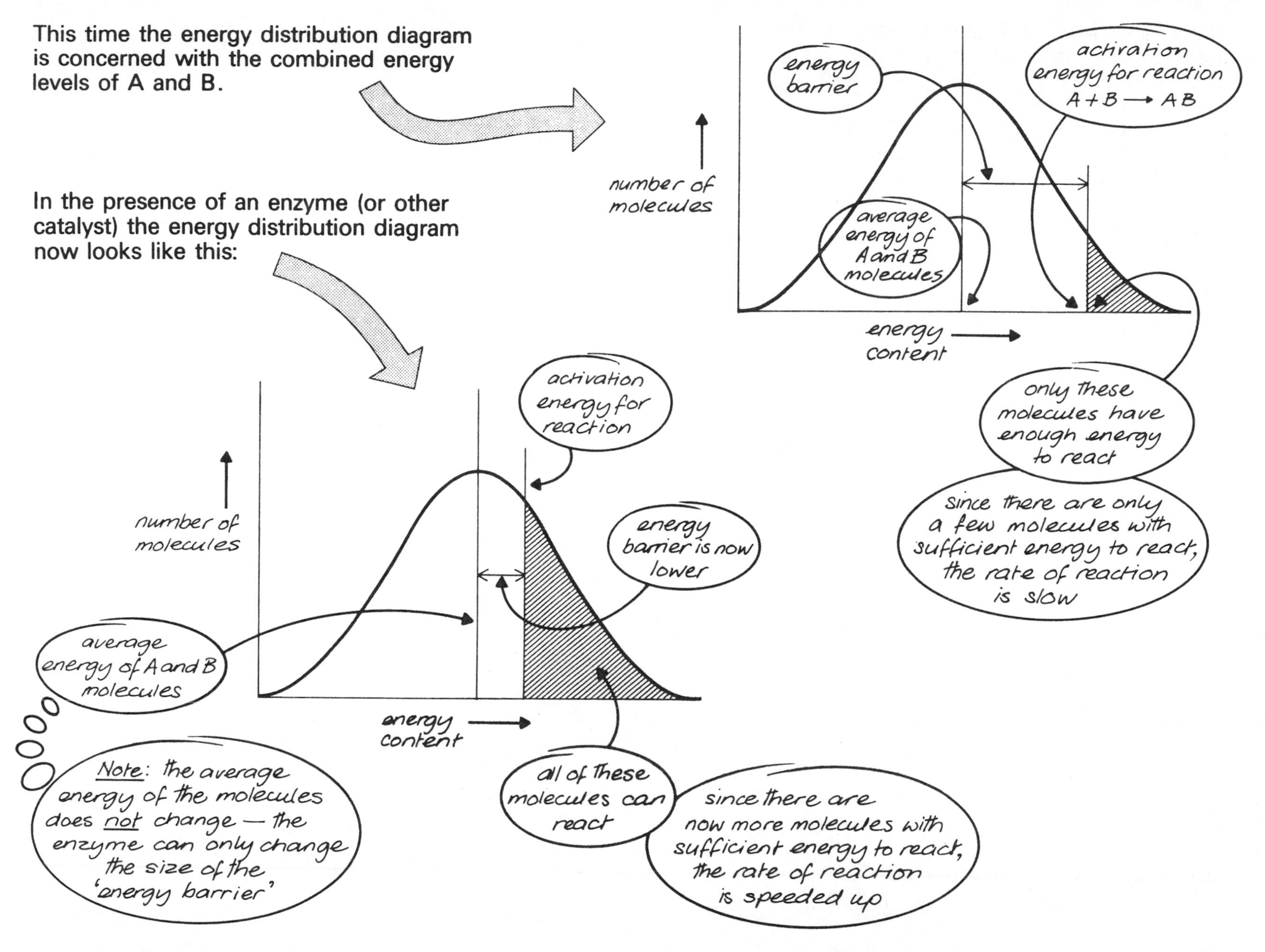

energy barrier
activation energy for reaction A + B ⟶ AB
number of molecules
average energy of A and B molecules
energy content
only these molecules have enough energy to react
since there are only a few molecules with sufficient energy to react, the rate of reaction is slow
activation energy for reaction
number of molecules
energy barrier is now lower
average energy of A and B molecules
energy content
Note: the average energy of the molecules does not change — the enzyme can only change the size of the 'energy barrier'
all of these molecules can react
since there are now more molecules with sufficient energy to react, the rate of reaction is speeded up

Part 3 Metabolism Itself

Glucose is a carbohydrate. Its formula is $C_6H_{12}O_6$. Living things can respire it to carbon dioxide (CO_2) and water (H_2O) using oxygen (O_2). During this process energy is extracted from the glucose molecule for use in various ways, for instance **work** (by muscles), for driving **energy-requiring (endergonic) chemical reactions** like protein synthesis, or even **light production** (by firefly tails). How is this done**?**

The equation for respiration is often written as

$$C_6H_{12}O_6 + 6O_2 \longrightarrow 6CO_2 + 6H_2O + \textit{energy}.$$

Although this equation is a correct summary of the quantitative changes which take place [for each mole (gram molecule) of glucose (180 g) disappearing, 6 moles of oxygen (192 g) are taken up from the air and 6 moles each of carbon dioxide (264 g) and water (108 g) appear in the cell] **it is misleading when we consider the mechanism of the process**. Such an equation is said to be correct only **stoichiometrically**.

According to the equation, glucose is oxidised both by removal of hydrogen and by the addition of oxygen. Research has shown that the cell separates these two processes in an unexpected way.

Process 1

Removal of hydrogen

Enzymes pass the hydrogen from glucose (and also from some water molecules in the cell) to molecules of a co-enzyme which can be represented by the symbol NAD. It then becomes $NADH_2$:

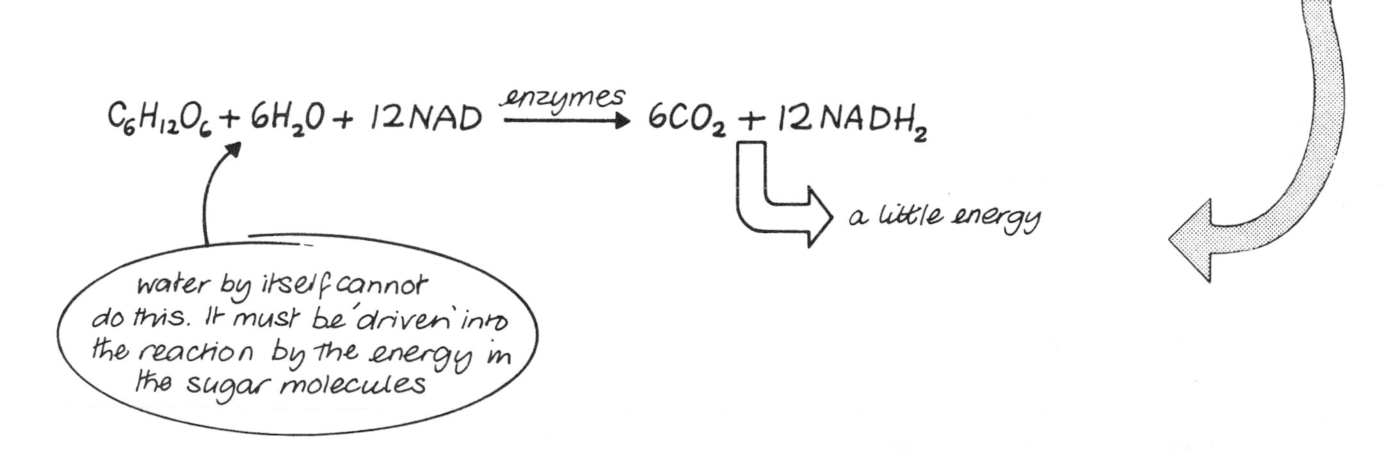
$C_6H_{12}O_6 + 6H_2O + 12NAD \xrightarrow{enzymes} 6CO_2 + 12NADH_2$
a little energy
water by itself cannot do this. It must be 'driven' into the reaction by the energy in the sugar molecules

All the CO_2 is released in this reaction, but only a little energy, most of the energy now being trapped in the $NADH_2$ molecules.

We can show the energy flow like this

Process 2

Addition of oxygen

Enzymes oxidise the $NADH_2$ using atmospheric oxygen, and generating water and NAD (the NAD is reused):

$$12\ NADH_2 + 6O_2 \xrightarrow{enzymes} 12H_2O + 12\ NAD + \text{much energy}$$

The rest of the energy is given up from the $NADH_2$ during this process.

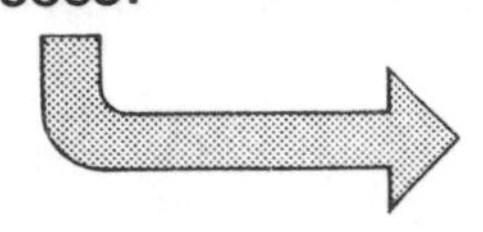

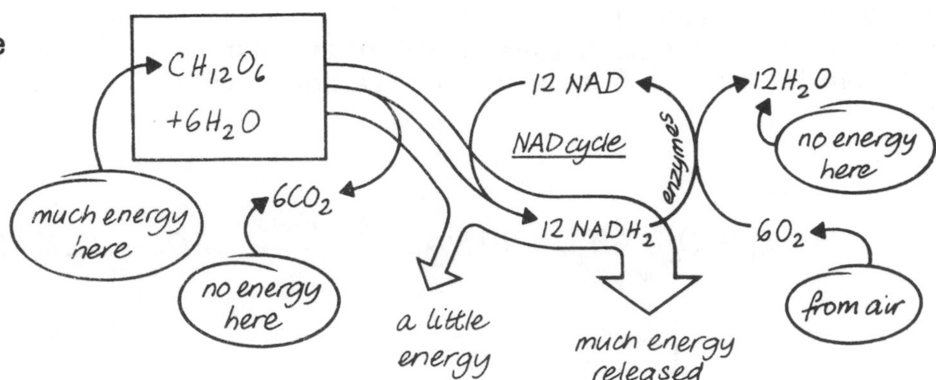

$CH_{12}O_6$
$+6H_2O$
$6CO_2$
much energy here
no energy here
a little energy
12 NAD
NAD cycle
enzymes
12 $NADH_2$
much energy released
$12H_2O$
no energy here
$6O_2$
from air

6 H_2O molecules enter the process (on the left) and 12 are generated (on the right), so the overall reaction, which can be summarised

$$C_6H_{12}O_6 + 6H_2O + 6O_2 \longrightarrow 6CO_2 + 12H_2O + \text{energy}$$

simplifies **stoichiometrically** to the equation for respiration

$$C_6H_{12}O_6 + 6O_2 \longrightarrow 6CO_2 + 6H_2O + \text{energy.}$$

Trapping the energy

The energy emerging from glucose oxidation (respiration) is trapped into another compound, called **adenosine triphosphate** symbolised by ATP.

ATP is quite complicated chemically, but we can write it simply as

Molecules of ATP can be looked upon as a battery of potential energy and are used to drive energy-requiring processes.

Note – The so-called 'high' energy bonds of ATP do not actually 'contain' higher energy: however **extra energy is released when bonds are broken**.

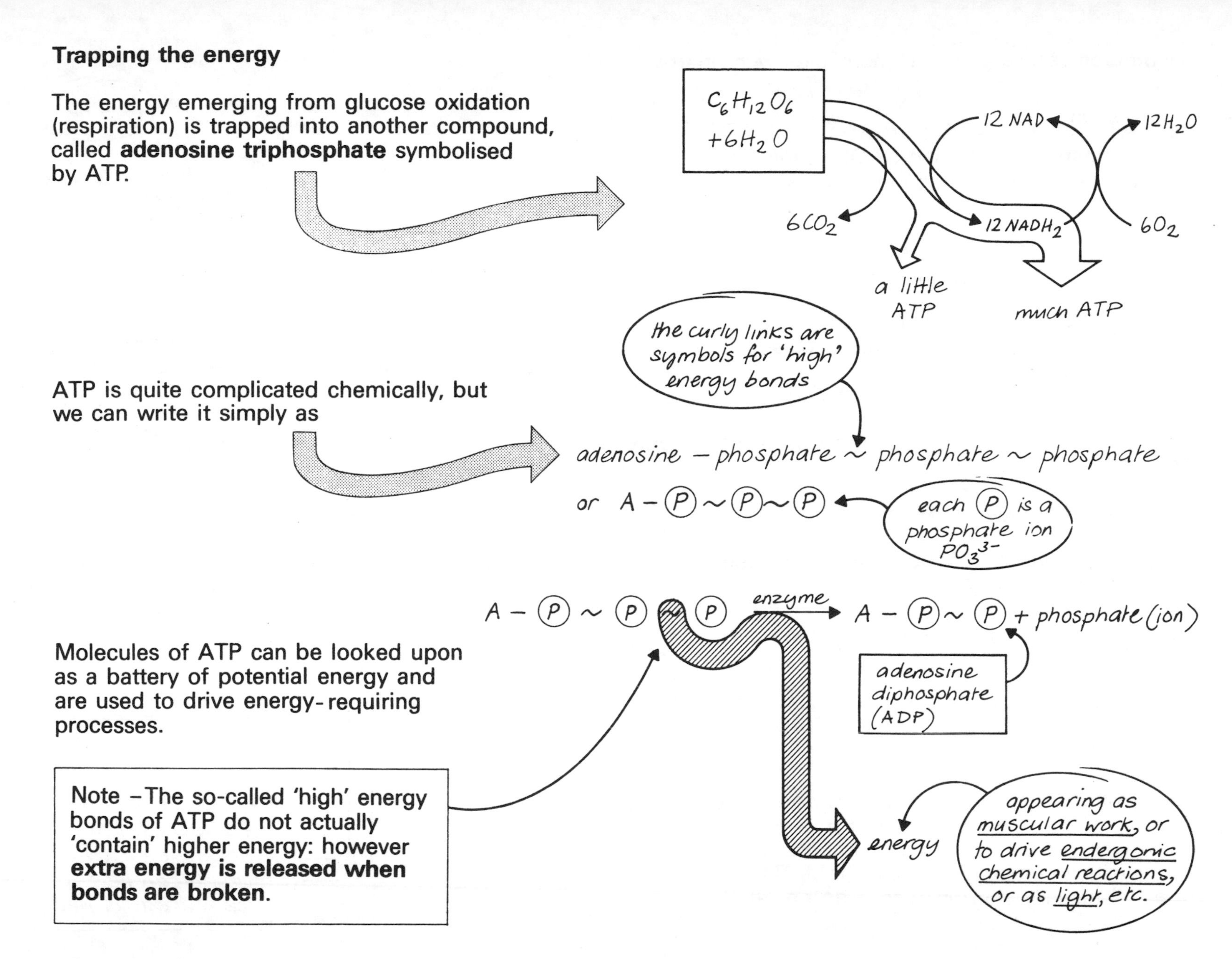

$C_6H_{12}O_6$ $+6H_2O$
12 NAD
$12H_2O$
$6CO_2$
12 $NADH_2$
$6O_2$
a little ATP
much ATP
the curly links are symbols for 'high' energy bonds
adenosine – phosphate ~ phosphate ~ phosphate
or A – (P) ~ (P) ~ (P)
each (P) is a phosphate ion PO_3^{3-}
A – (P) ~ (P) ~ (P)
enzyme
A – (P) ~ (P) + phosphate (ion)
adenosine diphosphate (ADP)
energy
appearing as muscular work, or to drive endergonic chemical reactions, or as light, etc.

Respiration re-charges ATP molecules with energy.

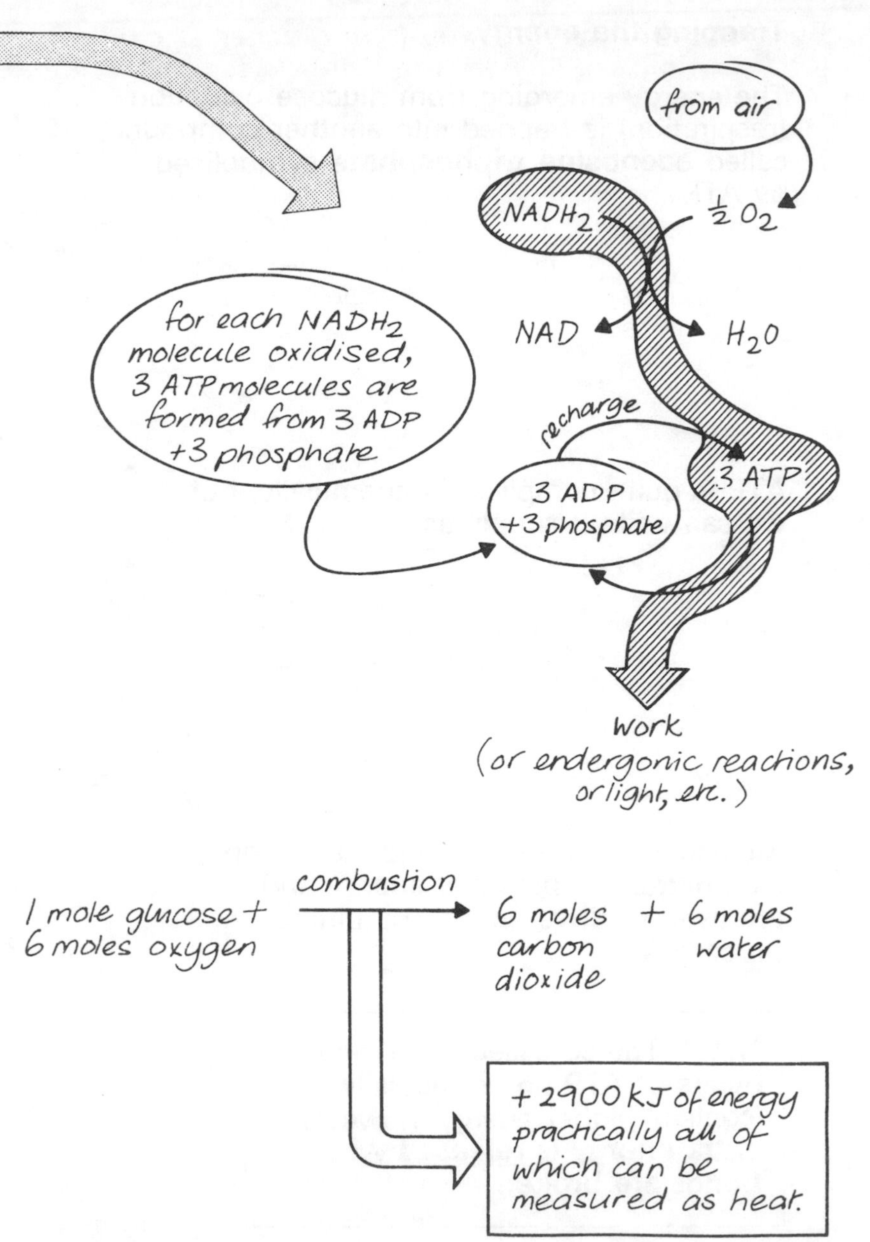

from air
$NADH_2$
$\frac{1}{2} O_2$
NAD
H_2O
for each $NADH_2$ molecule oxidised, 3 ATP molecules are formed from 3 ADP +3 phosphate
recharge
3 ATP
3 ADP +3 phosphate
work (or endergonic reactions, or light, etc.)
combustion
1 mole glucose + 6 moles oxygen
6 moles carbon dioxide
+ 6 moles water
+ 2900 kJ of energy practically all of which can be measured as heat.

As 3 ATP molecules are formed (from 3 ADP and 3 phosphate) for each $NADH_2$ molecule oxidised (to NAD and H_2O) we can calculate that since

12 $NADH_2$ molecules are formed for each glucose molecule oxidised

12 × 3 ATP molecules are formed for each glucose molecule oxidised

= **36 ATP.**

How much of the energy of the glucose molecule does this represent?

Or put another way:

Is all of the energy of the glucose molecule available for work?

If glucose is simply burnt in oxygen we get this stoichiometry:

But
the energy of glucose oxidation need not appear as heat. Some, or all of it, is capable of doing work or driving endergonic processes. This available energy is called **'Gibbs free energy'** (usually shortened to **'free energy'**).

By whatever pathway a mole of glucose is oxidised to carbon dioxide and water, it must lose this amount of free energy – no more, no less,

and it follows that a mole of glucose contains 2900 kJ **more** free energy than 6 moles of water plus 6 moles of carbon dioxide.

Similar equations can be written for the other compounds involved in respiration.

For instance

$$1 \text{ mole } NADH_2 + \tfrac{1}{2} \text{ mole } O_2 \longrightarrow 1 \text{ mole } NAD + 1 \text{ mole } H_2O$$

220 kJ free energy

When hydrogen atoms are transported from $NADH_2$ to oxygen they lose the free energy which is associated with them, and it becomes available for work.

We can express this idea in a general way by a diagram:

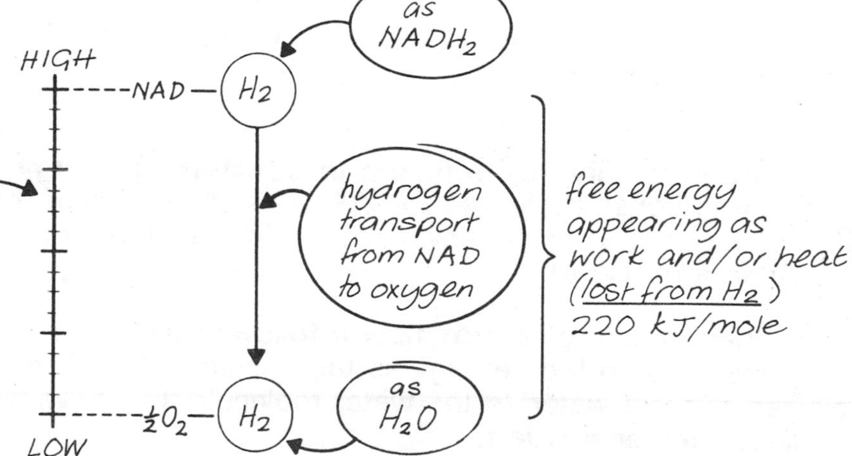

Scale of energy content of pairs of hydrogen atoms (kJ per mole)
HIGH
LOW
NAD
H2
as NADH2
hydrogen transport from NAD to oxygen
½O2
H2
as H2O
free energy appearing as work and/or heat (lost from H2) 220 kJ/mole

12 moles of $NADH_2$ contain 12 × 220 = 2640 kJ more free energy than 12 moles NAD + 12 moles H_2.

For the 12 moles of $NADH_2$ generated from 1 mole of glucose then

$\frac{2640}{2900} \times 100$ = about 90% of the energy of the glucose molecule is contained in the molecules of $NADH_2$.

Now consider ATP. The equation for the release of free energy from this compound is

ATP → ADP + phosphate

about 30 kJ free energy per mole

Note It follows that the formation of ATP from ADP (adenosine diphosphate) and phosphate ions **requires** an equivalent quantity of free energy.

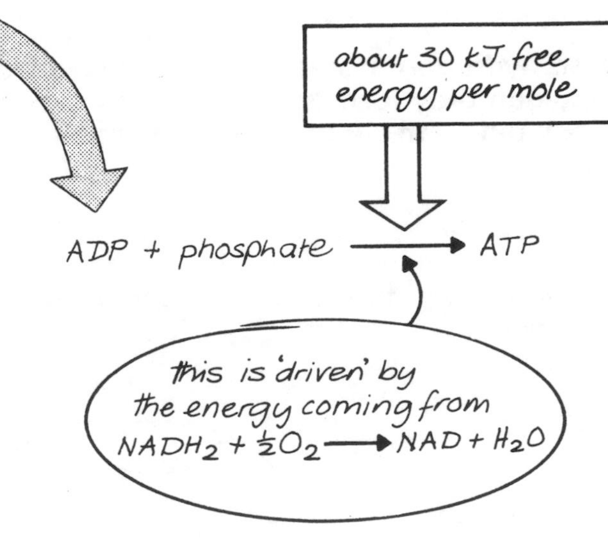
about 30 kJ free energy per mole
ADP + phosphate → ATP
this is 'driven' by the energy coming from $NADH_2 + \frac{1}{2}O_2 \rightarrow NAD + H_2O$

For the 36 moles of ATP regenerated (from ADP + phosphate) during the oxidation of 12 moles of $NADH_2$

36 × 30 = 1080 kJ of energy is required

which is $\frac{1080}{2900} \times 100$ = about 40% of the free energy of the glucose molecule.

Respiration is only 40% efficient in transferring energy from the glucose molecule to ATP. The other 60% of the free energy is lost to the environment ('wasted') in an unusable form (heat).

From this account of energy flow it follows that the pairs of H atoms yield up their energy as they become linked to oxygen to form water. In the water molecule they have no available free energy left.

More about respiration:

glycolysis, the Krebs cycle, oxidative phosphorylation.

We can now look at respiration in a little more detail, and include not only the path of energy and of hydrogen, but also the path of carbon leading to CO_2.

Biochemists have divided respiration into three subpathways.

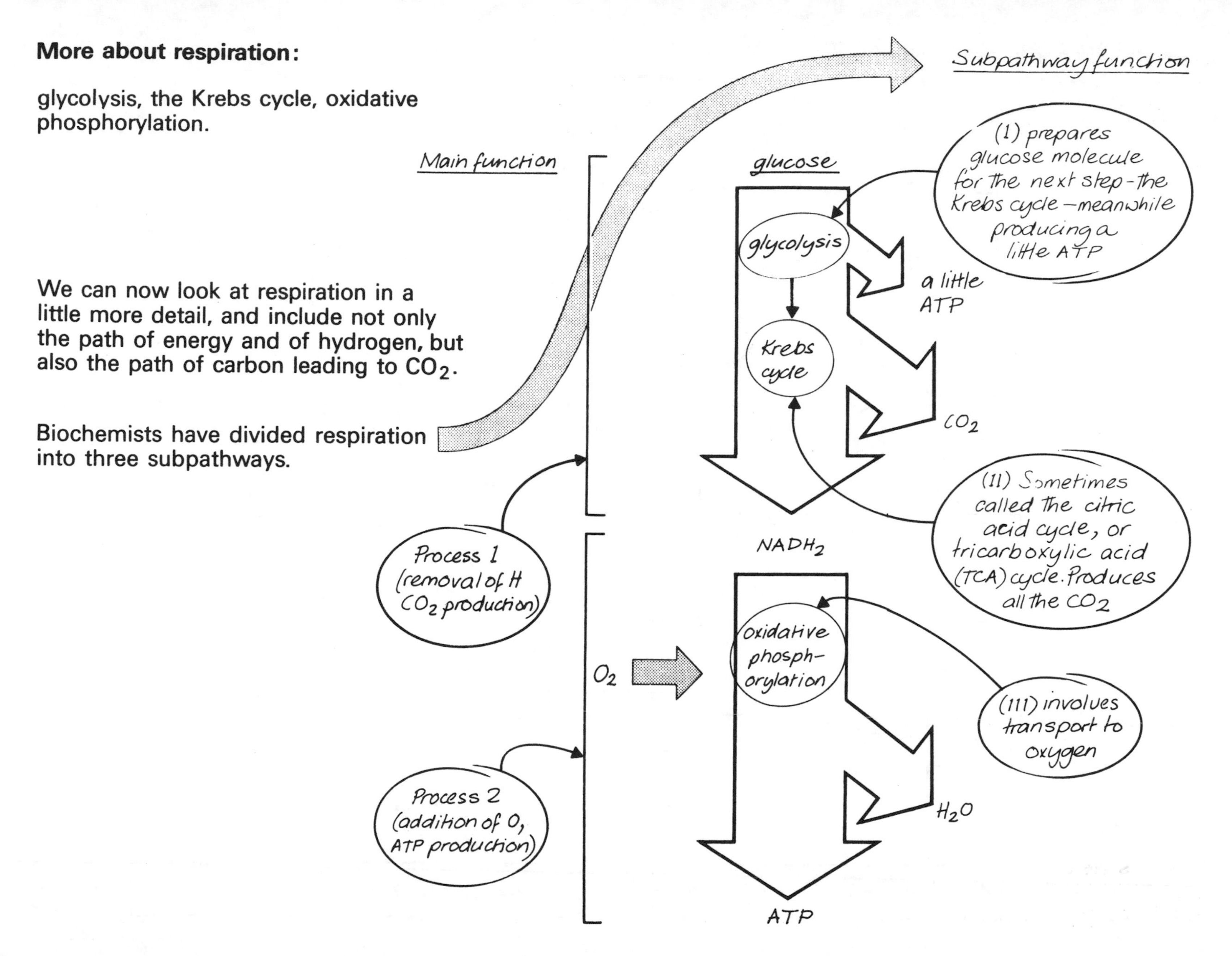

Subpathway function
Main function
glucose
glycolysis
(1) prepares glucose molecule for the next step - the Krebs cycle - meanwhile producing a little ATP
a little ATP
Krebs cycle
CO2
(11) Sometimes called the citric acid cycle, or tricarboxylic acid (TCA) cycle. Produces all the CO2
Process 1 (removal of H CO2 production)
NADH2
oxidative phosph-orylation
O2
(111) involves transport to oxygen
Process 2 (addition of O, ATP production)
H2O
ATP

More details about the subpathways

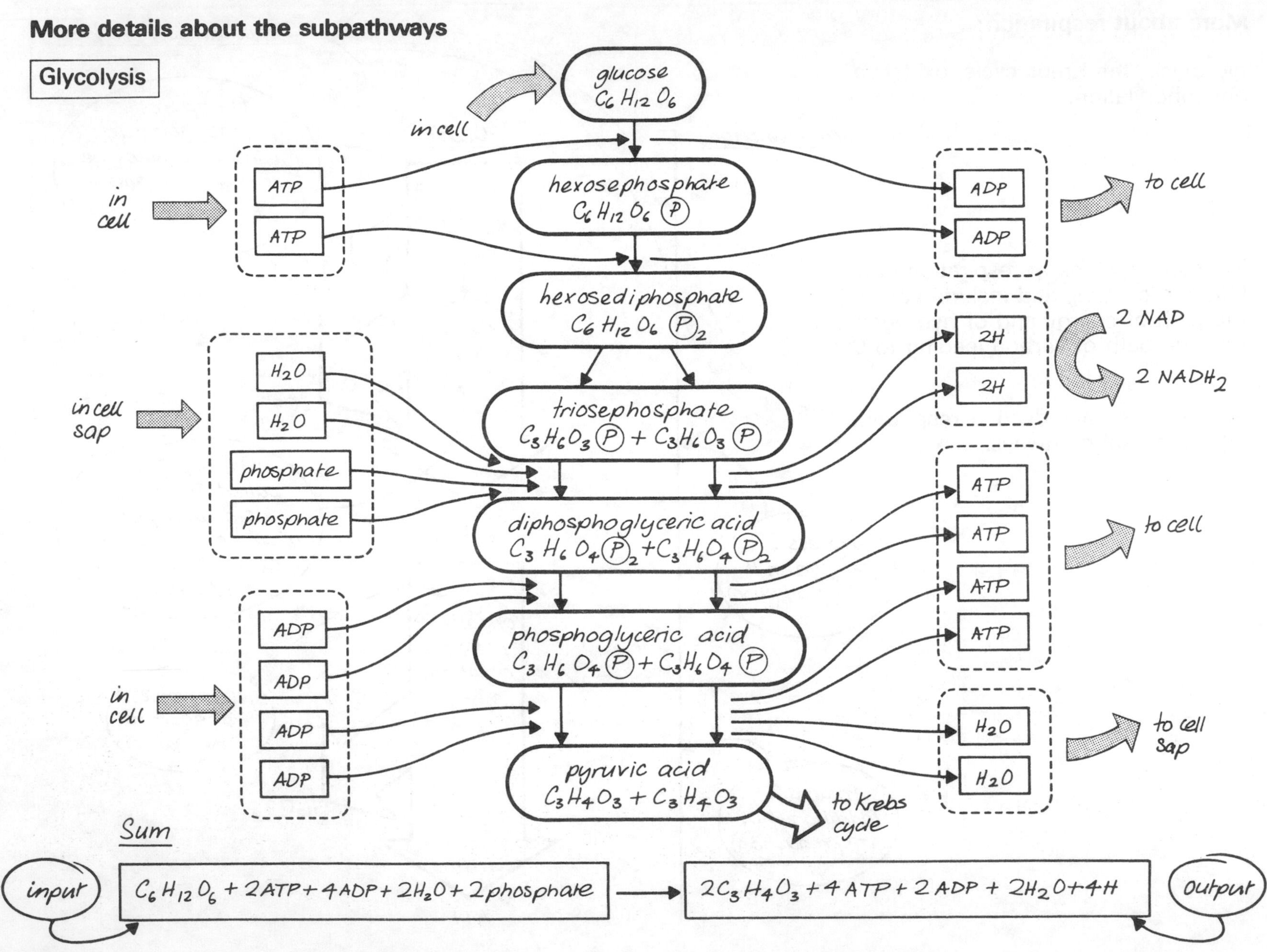

Glycolysis
in cell
glucose
$C_6H_{12}O_6$
in cell
ATP
ATP
hexosephosphate
$C_6H_{12}O_6$ (P)
ADP
ADP
to cell
hexosediphosphate
$C_6H_{12}O_6$ (P)$_2$
2H
2H
2 NAD
2 $NADH_2$
in cell sap
H_2O
H_2O
phosphate
phosphate
triosephosphate
$C_3H_6O_3$ (P) + $C_3H_6O_3$ (P)
diphosphoglyceric acid
$C_3H_6O_4$ (P)$_2$ + $C_3H_6O_4$ (P)$_2$
ATP
ATP
ATP
ATP
to cell
in cell
ADP
ADP
ADP
ADP
phosphoglyceric acid
$C_3H_6O_4$ (P) + $C_3H_6O_4$ (P)
H_2O
H_2O
to cell sap
pyruvic acid
$C_3H_4O_3 + C_3H_4O_3$
to Krebs cycle
Sum
input
$C_6H_{12}O_6$ + 2ATP + 4ADP + $2H_2O$ + 2 phosphate
$2C_3H_4O_3$ + 4 ATP + 2 ADP + $2H_2O$ + 4H
output

Pyruvic acid is now ready for further processing by the Krebs cycle, which converts it to a mixture of CO_2 and $NADH_2$ (page 65)

But–
If oxygen is not available for the subsequent oxidation of $NADH_2$ to water and NAD, the Krebs cycle will not function and the pyruvic acid is switched to other pathways: this happens when conditions are **anaerobic**.

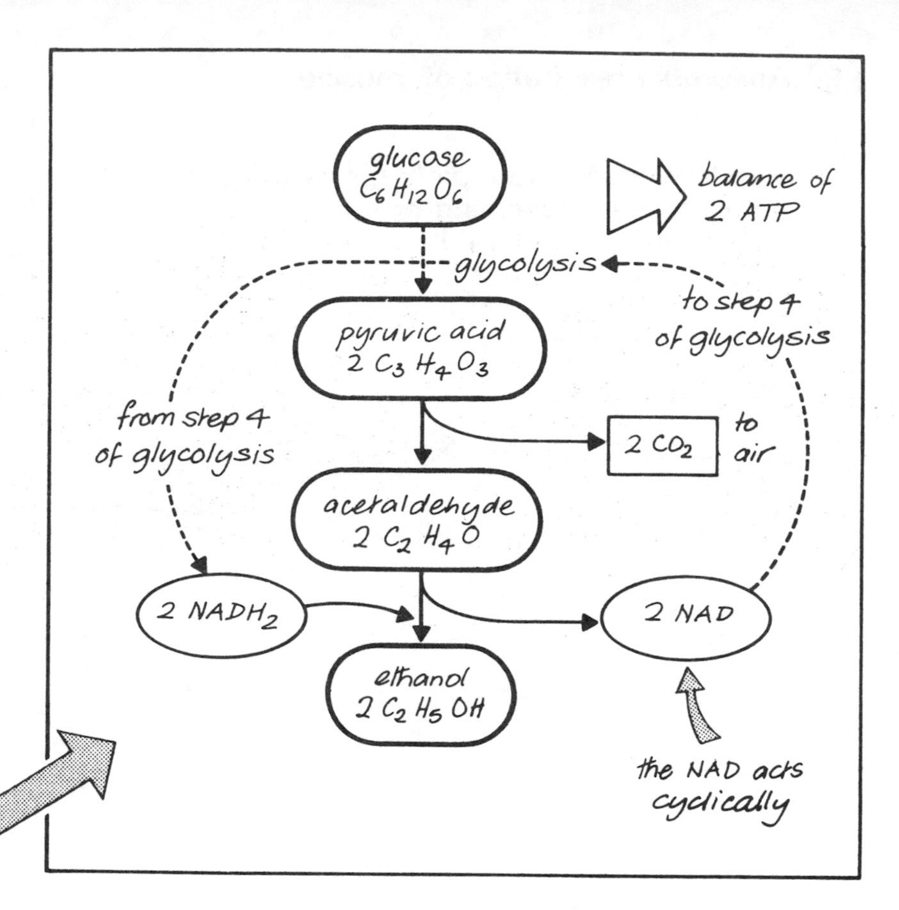
glucose
$C_6H_{12}O_6$
balance of
2 ATP
glycolysis
to step 4
of glycolysis
pyruvic acid
$2\ C_3H_4O_3$
from step 4
of glycolysis
2 CO_2
to
air
acetaldehyde
$2\ C_2H_4O$
2 $NADH_2$
2 NAD
ethanol
$2\ C_2H_5OH$
the NAD acts
cyclically

Two anaerobic cases:

Ⓐ **Fermentation (anaerobic metabolism) of sugar by yeast.**

Yeast growing in sugar solution rapidly depletes its oxygen supply. So unless vigorously stirred in air it will metabolise sugar like this

The stoichiometry of fermentation is:

$$C_6H_{12}O_6 + 2ADP + 2\ \text{phosphate} \longrightarrow 2C_2H_5OH + 2CO_2 + 2ATP$$

The **ATP** is used for cellular processes (work).
The ethanol is a byproduct. No oxygen is used.

(B) Anaerobic respiration of muscle

Muscle working too rapidly for its oxygen supply (e.g. in swimming or running) will metabolise sugar like this:

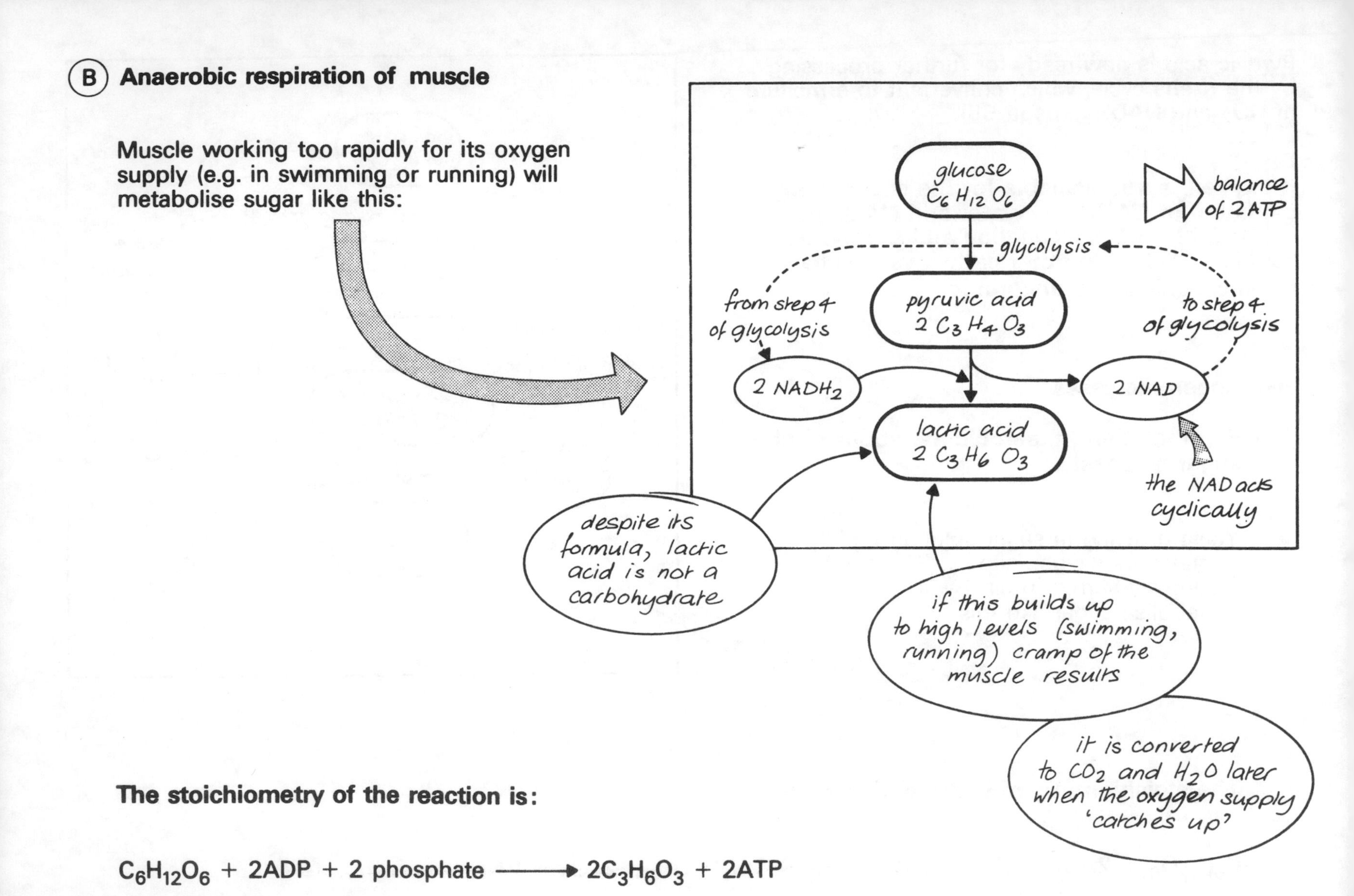

glucose $C_6H_{12}O_6$
balance of 2 ATP
glycolysis
from step 4 of glycolysis
pyruvic acid $2\,C_3H_4O_3$
to step 4 of glycolysis
2 $NADH_2$
2 NAD
lactic acid $2\,C_3H_6O_3$
the NAD acts cyclically
despite its formula, lactic acid is not a carbohydrate
if this builds up to high levels (swimming, running) cramp of the muscle results
it is converted to CO_2 and H_2O later when the oxygen supply 'catches up'

The stoichiometry of the reaction is:

$$C_6H_{12}O_6 + 2ADP + 2\text{ phosphate} \longrightarrow 2C_3H_6O_3 + 2ATP$$

The Krebs cycle

Converts pyruvic acid from glycolysis entirely to CO_2, generating $NADH_2$ in the process.

It is cyclic because one of the later products (oxaloacetic acid) is used as a 'carrier' to introduce more pyruvic acid into the pathway.

We can summarise the **principle** of the cycle like this:

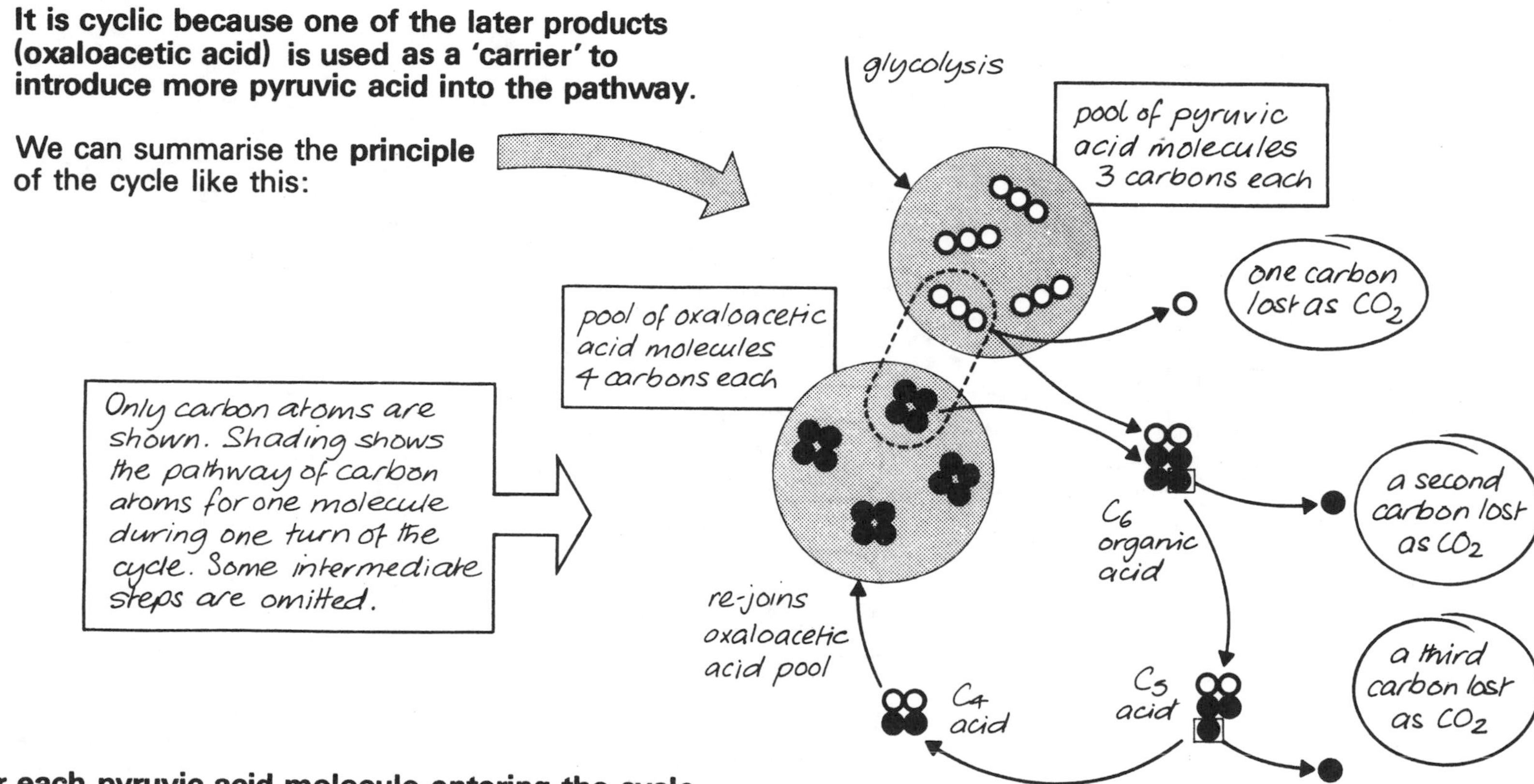

glycolysis
pool of pyruvic acid molecules 3 carbons each
one carbon lost as CO2
pool of oxaloacetic acid molecules 4 carbons each
Only carbon atoms are shown. Shading shows the pathway of carbon atoms for one molecule during one turn of the cycle. Some intermediate steps are omitted.
a second carbon lost as CO2
C6 organic acid
re-joins oxaloacetic acid pool
a third carbon lost as CO2
C5 acid
C4 acid

For each pyruvic acid molecule entering the cycle, $3CO_2$ molecules are produced so the stoichiometry for carbon is correct. Once again, however, **the stoichiometry tells us nothing of the mechanism.** The carbon atoms from each pyruvate emerge, not specifically as it goes round the cycle, but during later turns in the cycle (you can see this if you follow an oxaloacetic acid molecule around two or three times).

We can now give the cycle in a little more detail.
Each of the carbon intermediates is an acid.

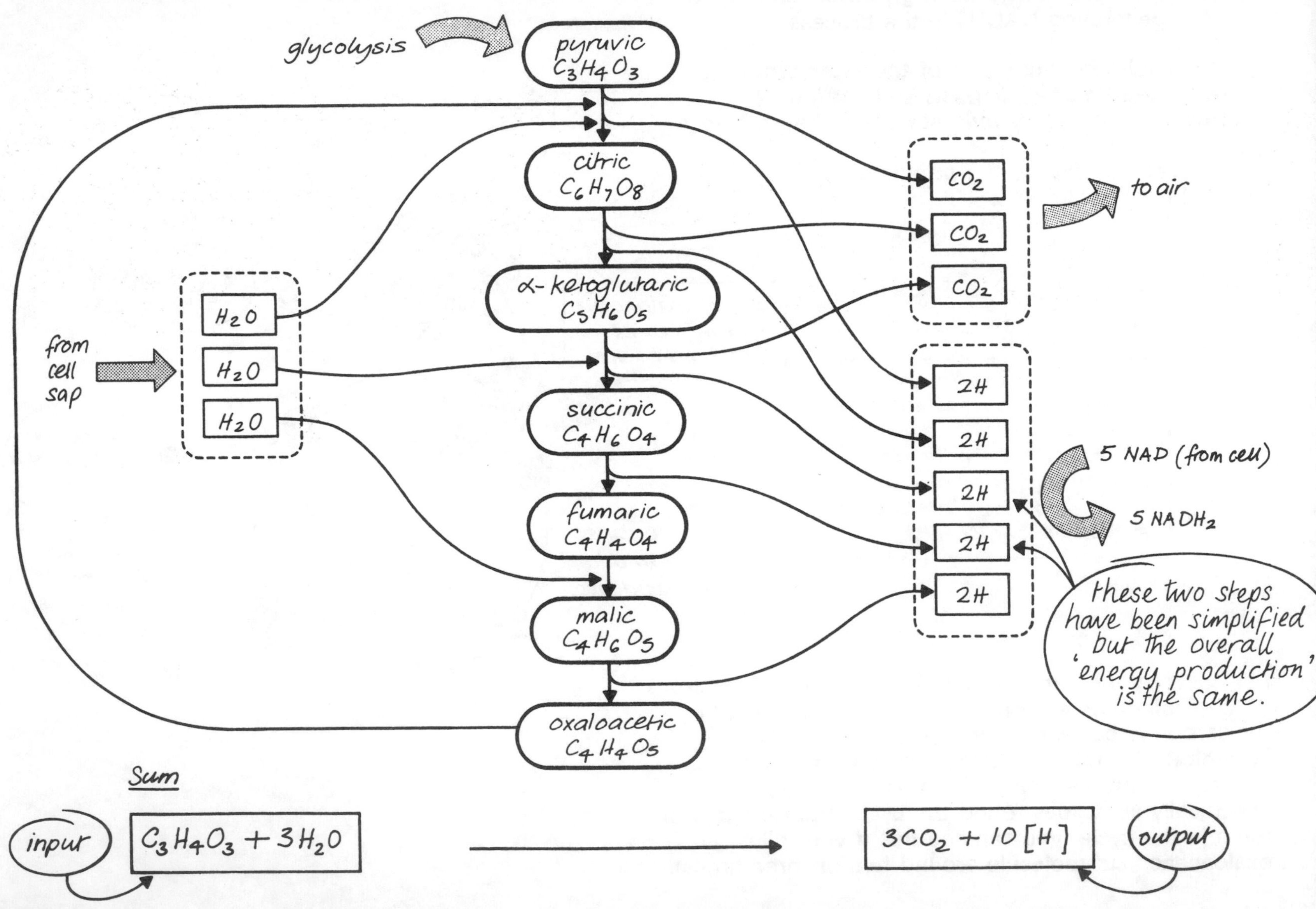

glycolysis
pyruvic $C_3H_4O_3$
citric $C_6H_7O_8$
α-ketoglutaric $C_5H_6O_5$
succinic $C_4H_6O_4$
fumaric $C_4H_4O_4$
malic $C_4H_6O_5$
oxaloacetic $C_4H_4O_5$
from cell sap
H_2O
H_2O
H_2O
CO_2
CO_2
CO_2
to air
2H
2H
2H
2H
2H
5 NAD (from cell)
5 $NADH_2$
these two steps have been simplified but the overall 'energy production' is the same.
Sum
input
$C_3H_4O_3 + 3H_2O$
$3CO_2 + 10[H]$
output

Each glucose molecule gives 2 pyruvic acid molecules by glycolysis, so we can summarise both processes as:

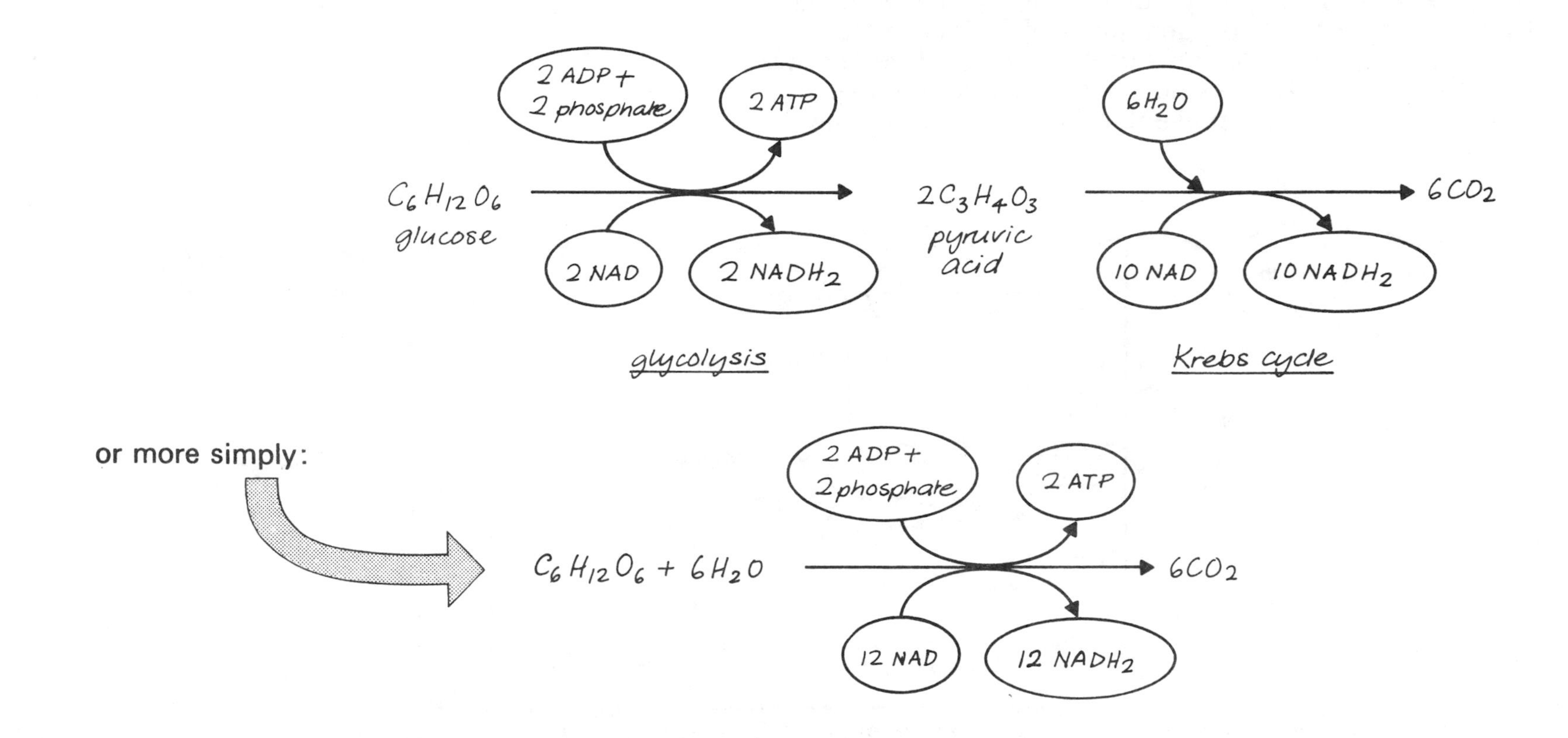

2 ADP + 2 phosphate
2 ATP
$C_6H_{12}O_6$ glucose
2 NAD
2 $NADH_2$
$2C_3H_4O_3$ pyruvic acid
$6H_2O$
$6CO_2$
10 NAD
10 $NADH_2$
glycolysis
Krebs cycle
or more simply:
2 ADP + 2 phosphate
2 ATP
$C_6H_{12}O_6 + 6H_2O$
$6CO_2$
12 NAD
12 $NADH_2$

Oxidative Phosphorylation

During this process, the $NADH_2$ is oxidised by oxygen to NAD (which is re-used in the Krebs cycle) and water

$$12\ NADH_2 + 6\ O_2 \longrightarrow 12\ NAD + 12\ H_2O$$

At the same time ATP is generated.

How is this achieved?

We have already seen that the transport of hydrogen from $NADH_2$ to oxygen, yielding water, is accompanied by a loss from the hydrogen pairs of 220 kJ/mole, this appearing as potentially usable energy (free energy). This transport of H does not take place in one step. The process can be represented in the following way:

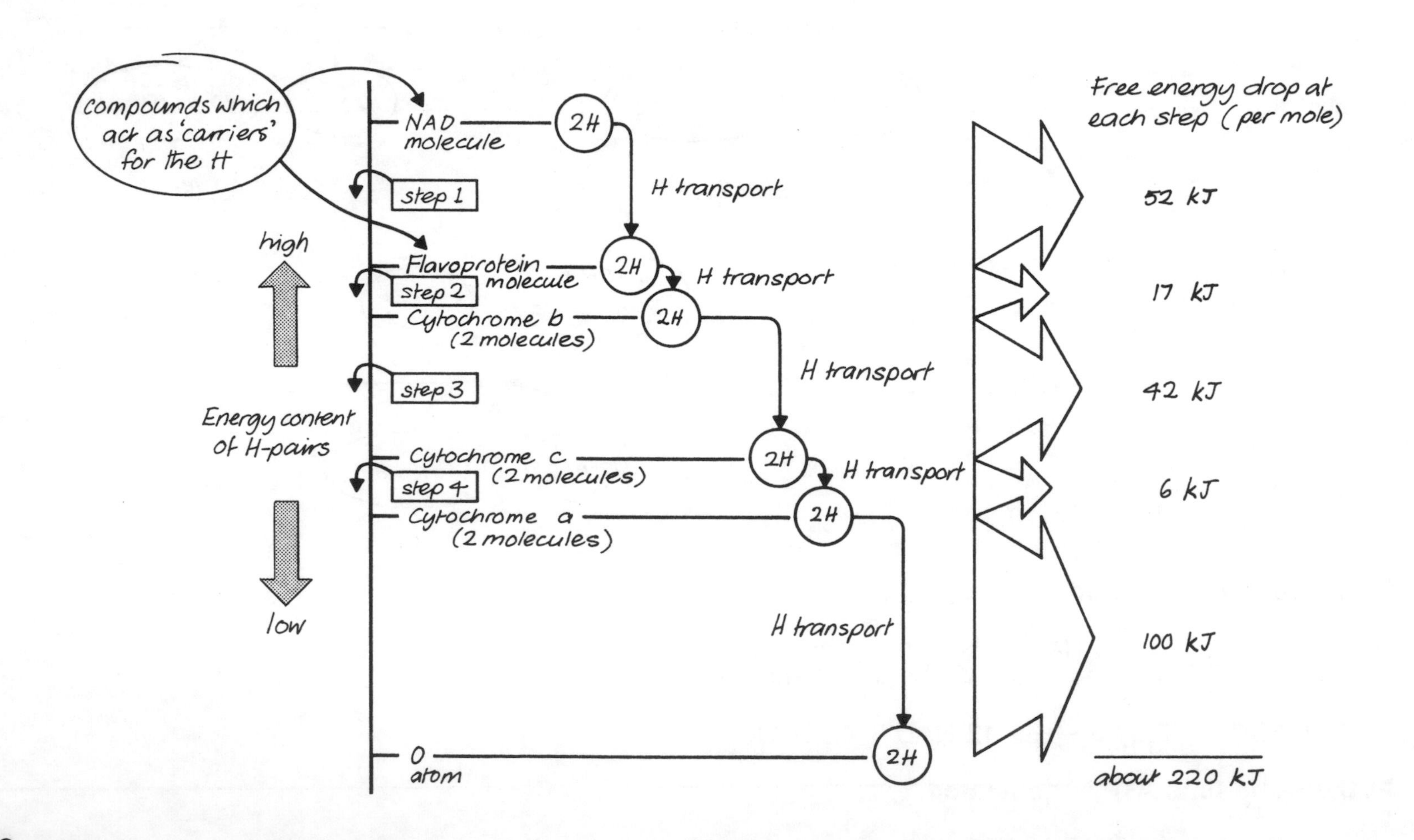

compounds which act as 'carriers' for the H
NAD molecule
2H
H transport
step 1
high
Flavoprotein molecule
step 2
Cytochrome b (2 molecules)
step 3
Energy content of H-pairs
Cytochrome c (2 molecules)
step 4
Cytochrome a (2 molecules)
low
O atom
Free energy drop at each step (per mole)
52 kJ
17 kJ
42 kJ
6 kJ
100 kJ
about 220 kJ

A conceptual diversion

This concept is easier to grasp if we use the analogy of a ball rolling down some steps not necessarily all of the same height.

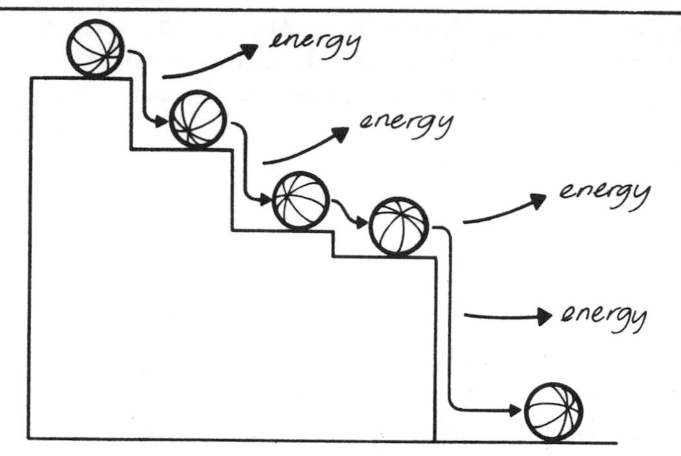
energy
energy
energy
energy
bottom of steps or 'ground level'

The ball loses potential energy at each step, the amount depending on the length of the vertical fall.

This energy can be used:
e.g. for one step

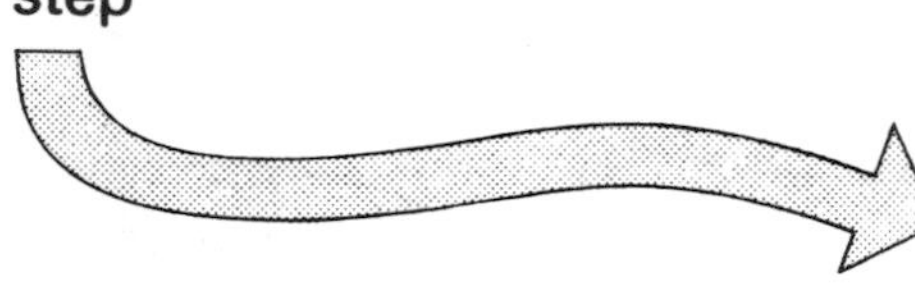

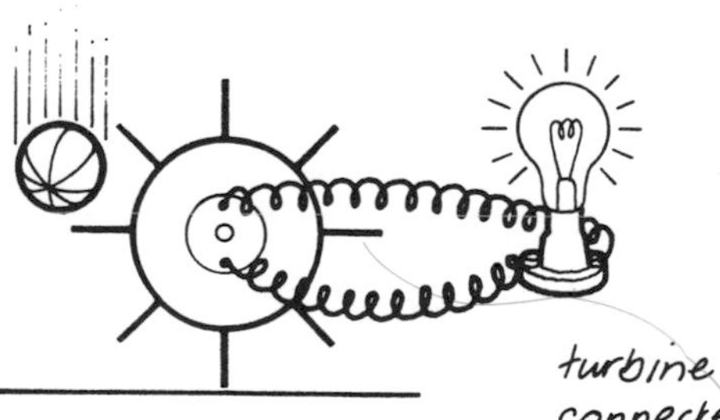
crankshaft connected to working machine
turbine connected to electric generator

Note that the **absolute** heights of the top and bottom steps do not matter. It is the **difference in height** which is important – 'ground level' for the system shown could be the top of a plateau or sea level, or the sea bed. In the same way the formation of water can be considered as 'reaching ground level' for the energy content of the H atoms as far as living organisms are concerned. Note also that not all of the potential energy (whole distance) of each fall need be used.

In the diagram there are three places where the energy loss from the H-pairs would be sufficient to drive the reaction:

30 kJ/mole

ADP + phosphate → ATP

One of them (the fifth) could theoretically generate more than one ATP molecule per H-pair but for reasons unknown, does not do so, and much of its free energy is lost to the environment (as heat).
The transfer of one hydrogen pair to an O atom therefore regenerates 3 ATP molecules.

$NADH_2 + \frac{1}{2}O_2$ → $NAD + H_2O$

3 ADP + 3 phosphate → 3 ATP

So for 12 $NADH_2$ molecules generated from one glucose molecule:

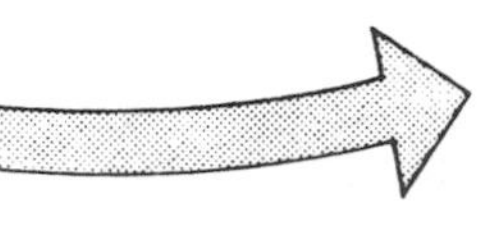

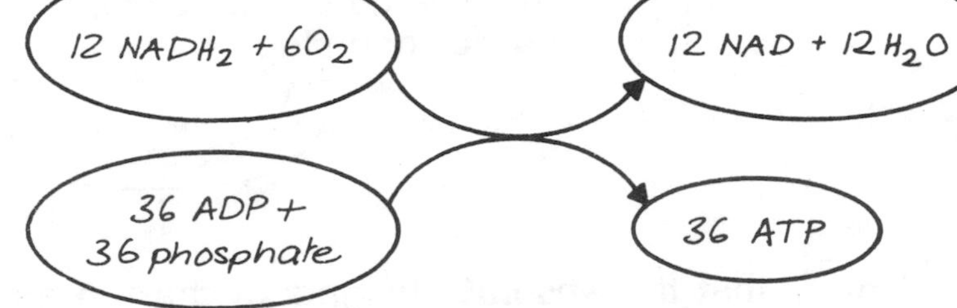
12 NADH2 + 6O2
12 NAD + 12 H2O
36 ADP + 36 phosphate
36 ATP

Or finally stoichiometrically:

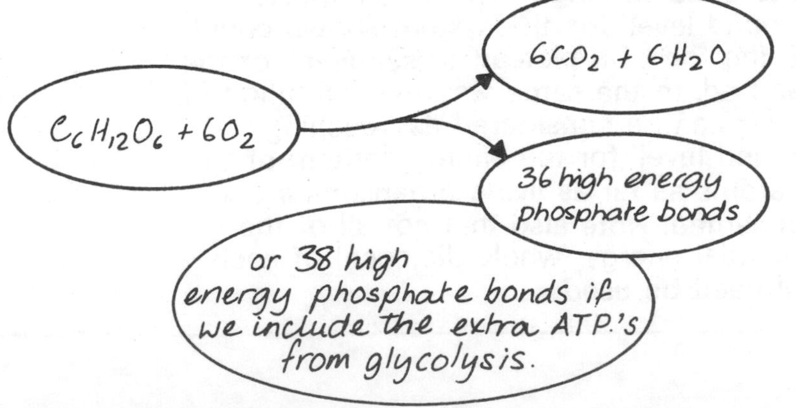
C6H12O6 + 6O2
6CO2 + 6H2O
36 high energy phosphate bonds
or 38 high energy phosphate bonds if we include the extra ATP's from glycolysis.

efficiency of energy transfer about 40%

A conceptual diversion

The steps we have called 'hydrogen transport' (page 68) are usually called **electron transport**. Why?

A knowledge of chemistry tells us that chemical reduction is not merely addition of hydrogen or removal of oxygen.

> **Reduction is basically the addition of electons.**
> **Oxidation is basically the removal of electrons.**

The hydrogen atom [H] can be considered to be a positively charged nucleus H^+ together with its electron e^-.

hydrogen ion — *electron*

$$[H] \equiv [H^+ + e^-]$$

The hydrogen ion H^+ is a hydrogen atom which has lost its electron.

When we add a hydrogen **atom** to a molecule we necessarily add an electron, and so carry out a reduction.

e.g. $S + H_2 \longrightarrow H_2S$ (*sulphur*) — *reduction; requires much energy*

When we add merely a hydrogen **ion** we do not carry out a reduction.

e.g. $S^{2-} + 2H^+ \longrightarrow H_2S$ (*sulphide*) — *no reduction; little energy change*

Now look at the diagram on page 68.

When an enzyme 'passes' the reducing power of $NADH_2$ on to flavoprotein it does so like this.

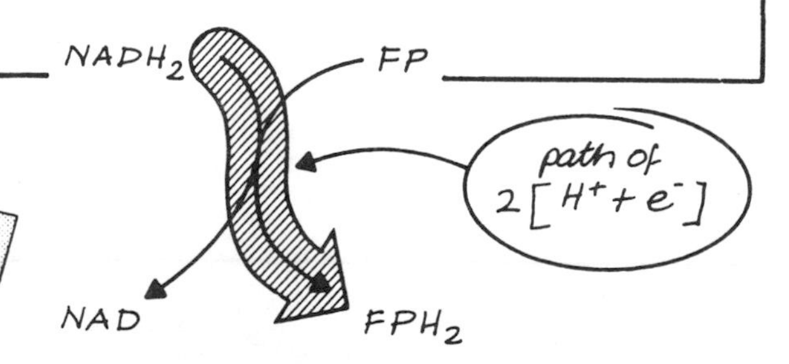
NADH2
FP
path of 2[H⁺+e⁻]
NAD
FPH2

But when FPH_2 has its reducing power passed on to cytochrome b **only the electrons are transported**, because cytochrome b (like all the cytochromes) contains an iron atom which can be either in the trivalent (more oxidised) state or in the divalent (more reduced) state

$$Fe^{3+} + e^- \longrightarrow Fe^{2+} \text{ reduction}$$

$$Fe^{2+} - e^- \longrightarrow Fe^{3+} \text{ oxidation}$$

So

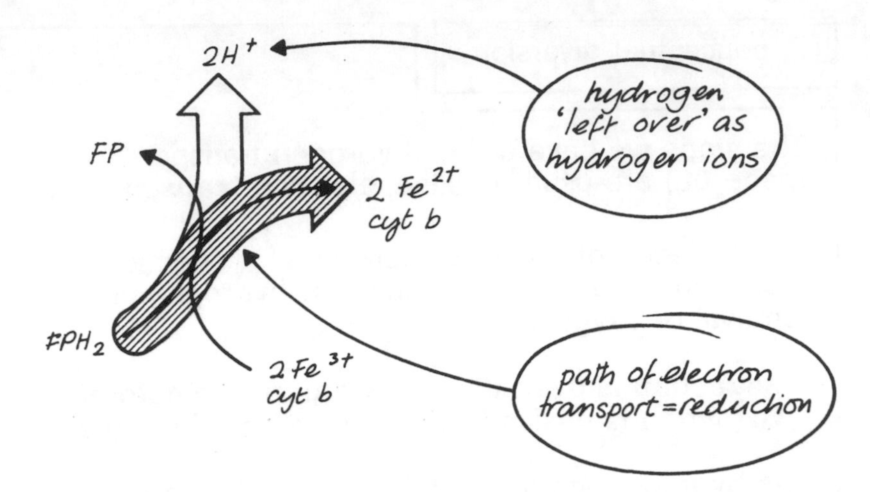

2H+
hydrogen 'left over' as hydrogen ions
FP
2 Fe2+ cyt b
FPH2
2 Fe3+ cyt b
path of electron transport = reduction

This is continued in the next steps until the electrons are passed to oxygen, reducing it to water.

So the **electron transport chain** becomes:

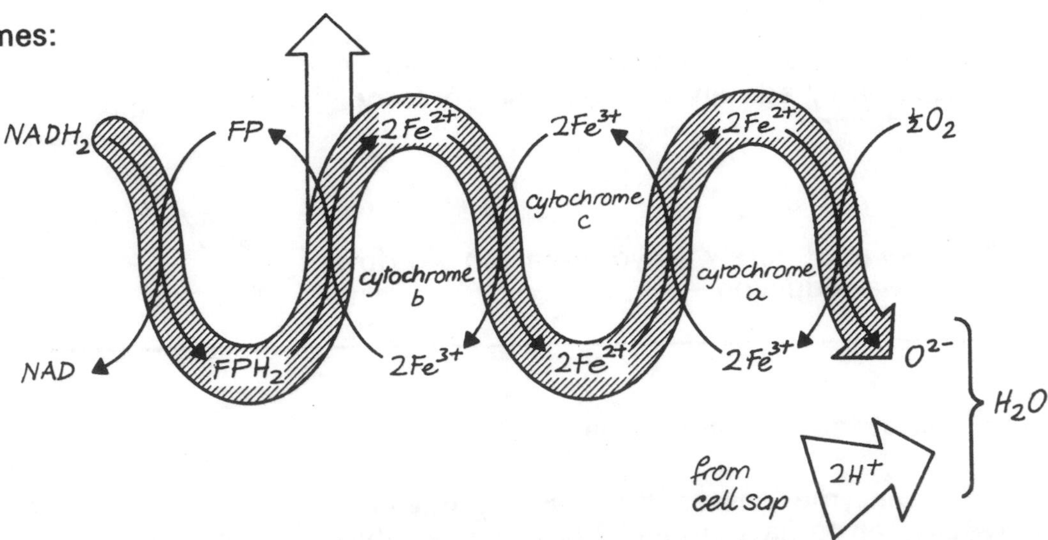

2H+
into cell sap
NADH2
FP
2Fe2+
2Fe3+
2Fe2+
½O2
cytochrome c
cytochrome b
cytochrome a
NAD
FPH2
2Fe3+
2Fe2+
2Fe3+
O2-
H2O
from cell sap
2H+

There are other ways of respiring carbohydrate in the living world.

The respiratory process already described (glycolysis ⟶ Krebs cycle ⟶ electron transport) is by far the most widespread and important one to be found in living cells but there are others (especially among the bacteria). Perhaps the next most important is the **pentose phosphate pathway** (or PPP).

This process also utilises glucose as the starting material.

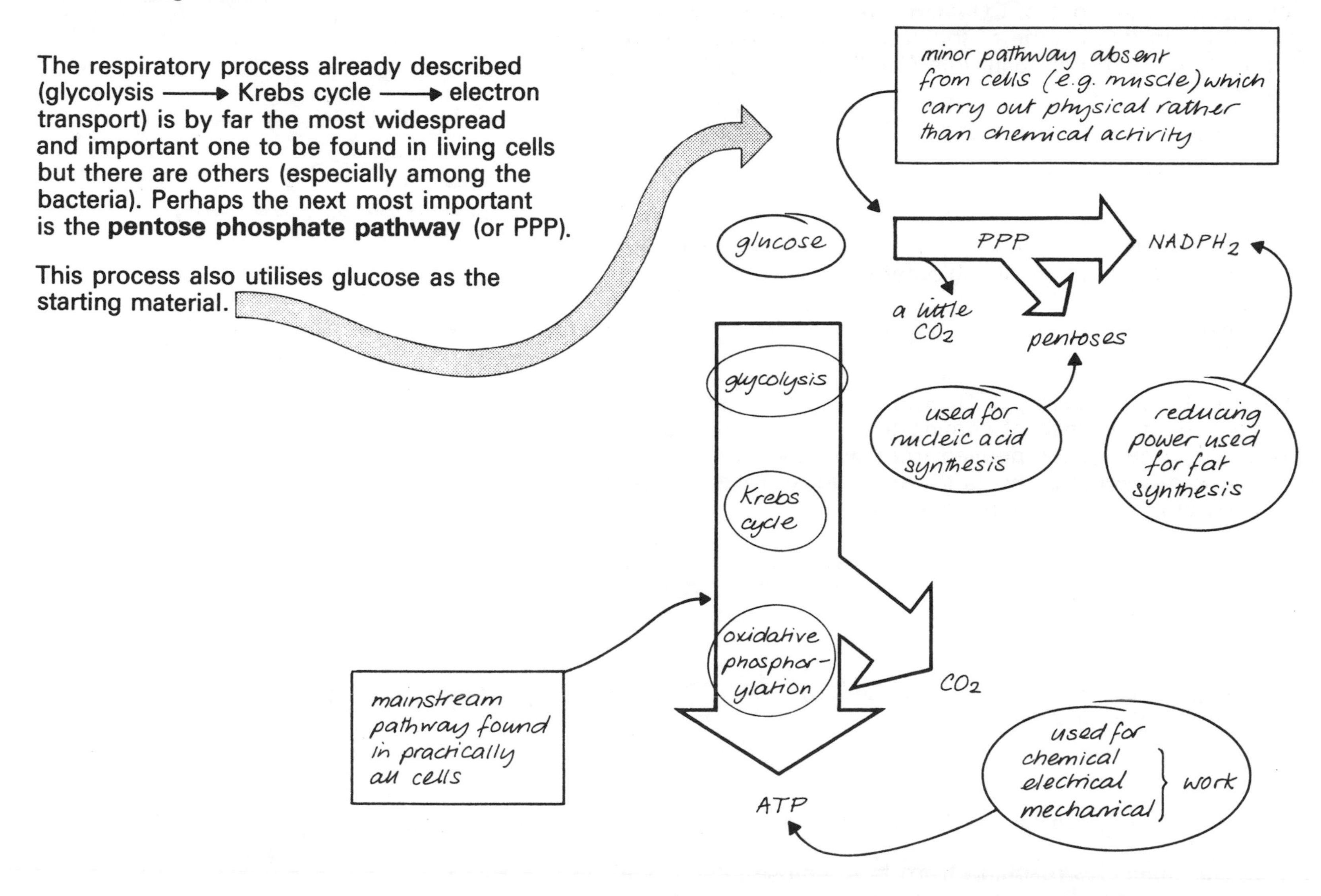
minor pathway absent from cells (e.g. muscle) which carry out physical rather than chemical activity
glucose
PPP
NADPH2
a little CO2
pentoses
glycolysis
used for nucleic acid synthesis
reducing power used for fat synthesis
Krebs cycle
oxidative phosphorylation
CO2
mainstream pathway found in practically all cells
ATP
used for chemical electrical mechanical work

In the pentose phosphate pathway, glucose (in the form of glucose phosphate) is partially oxidised by the removal of hydrogen which results in the loss of CO_2 and the formation of a pentose sugar. In this process some of the energy of glucose is transferred with the H atoms to the coenzyme **NADP** to form **$NADPH_2$**.

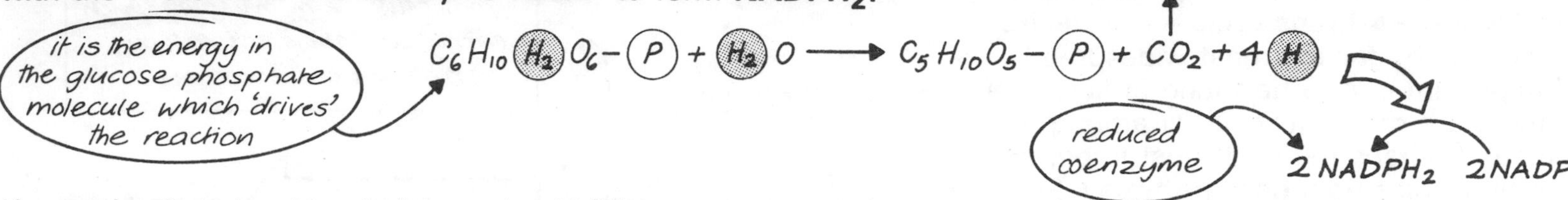

it is the energy in the glucose phosphate molecule which 'drives' the reaction
$C_6H_{10}H_2O_6-$(P) + H_2O ⟶ $C_5H_{10}O_5-$(P) + CO_2 + 4(H)
reduced coenzyme
2 $NADPH_2$
2NADP

If we take 6 molecules of glucose (and **billions** take part, even in this minor pathway) the stoichiometry is:

$$6C_6H_{12}O_6-\text{(P)} + 6H_2O + 12NADP \longrightarrow 6C_5H_{10}O_5-\text{(P)} + 6CO_2 + 12\,NADPH_2$$

(input) (output)

Looking at the output side, $6CO_2$ and $12NADPH_2$ have come from glucose, but instead of coming from a single glucose molecule, they have been derived from **one-sixth of each of six glucose molecules**. This is the **stoichiometric equivalent** to the oxidation of one glucose molecule.

We will see later (in the section on photosynthesis) that pentose phosphates and hexose phosphates can be interconverted. This happens readily in cells which contain the pentose phosphate pathway.

We have: ($6C_5 \equiv 5C_6$ stoichiometrically)

$$6C_5H_{10}O_5-\text{(P)} \xrightarrow{\text{many steps}} 5C_6H_{12}O_6-\text{(P)} + \text{(Pi)}$$

So we have:

$$6C_6H_{12}O_6-\text{(P)} + 6H_2O + 12NADP \longrightarrow 5C_6H_{12}O_6-\text{(P)} + 6CO_2 + 12NADPH_2 + \text{(Pi)}$$

The hexose molecules resulting from the pentose interconversions can of course be re-utilised in respiration.

Respiration of fats

Fats as well as carbohydrates can be respired to yield useful energy. The first step splits the molecules of triglycerides into glycerol and fatty acid molecules.

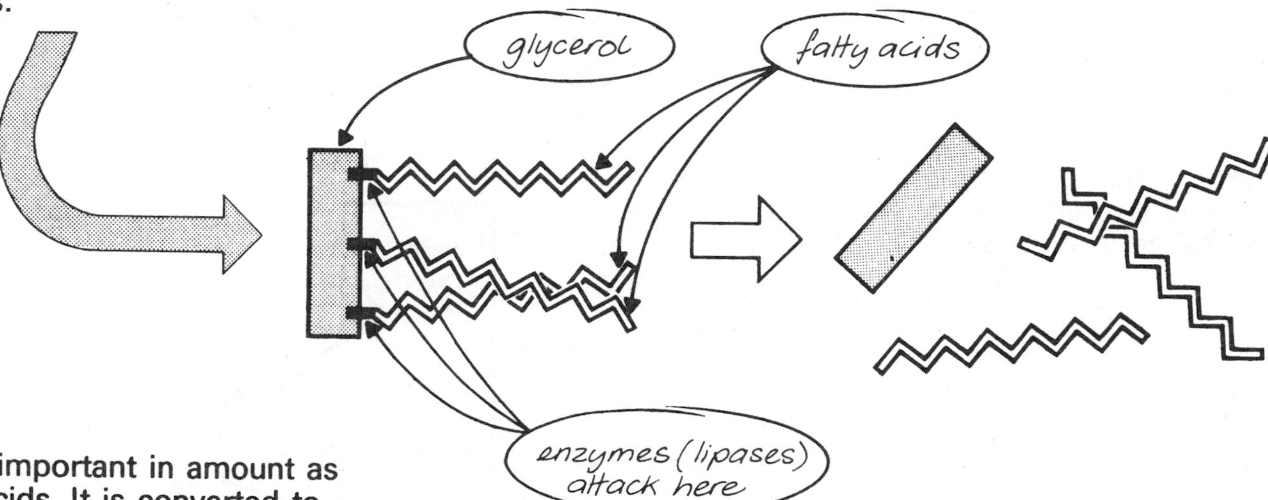

glycerol
fatty acids
enzymes (lipases) attack here

The glycerol is not very important in amount as compared to the fatty acids. It is converted to **triose phosphate** and then follows glycolysis etc. (see page 62) About 90% of the weight of a fat is fatty acid. Fatty acids contain a lot of energy. For example:

$$1 \text{ mole } C_{16}H_{32}O_2 + 23 \text{ moles } O_2 \xrightarrow{\text{combustion}} 16 \text{ moles } CO_2 + 16 \text{ moles } H_2O$$

10,000 kJ of energy

This is a stoichiometric equation (**remember**: it tells us nothing of the biochemical pathway of respiration of fatty acids).

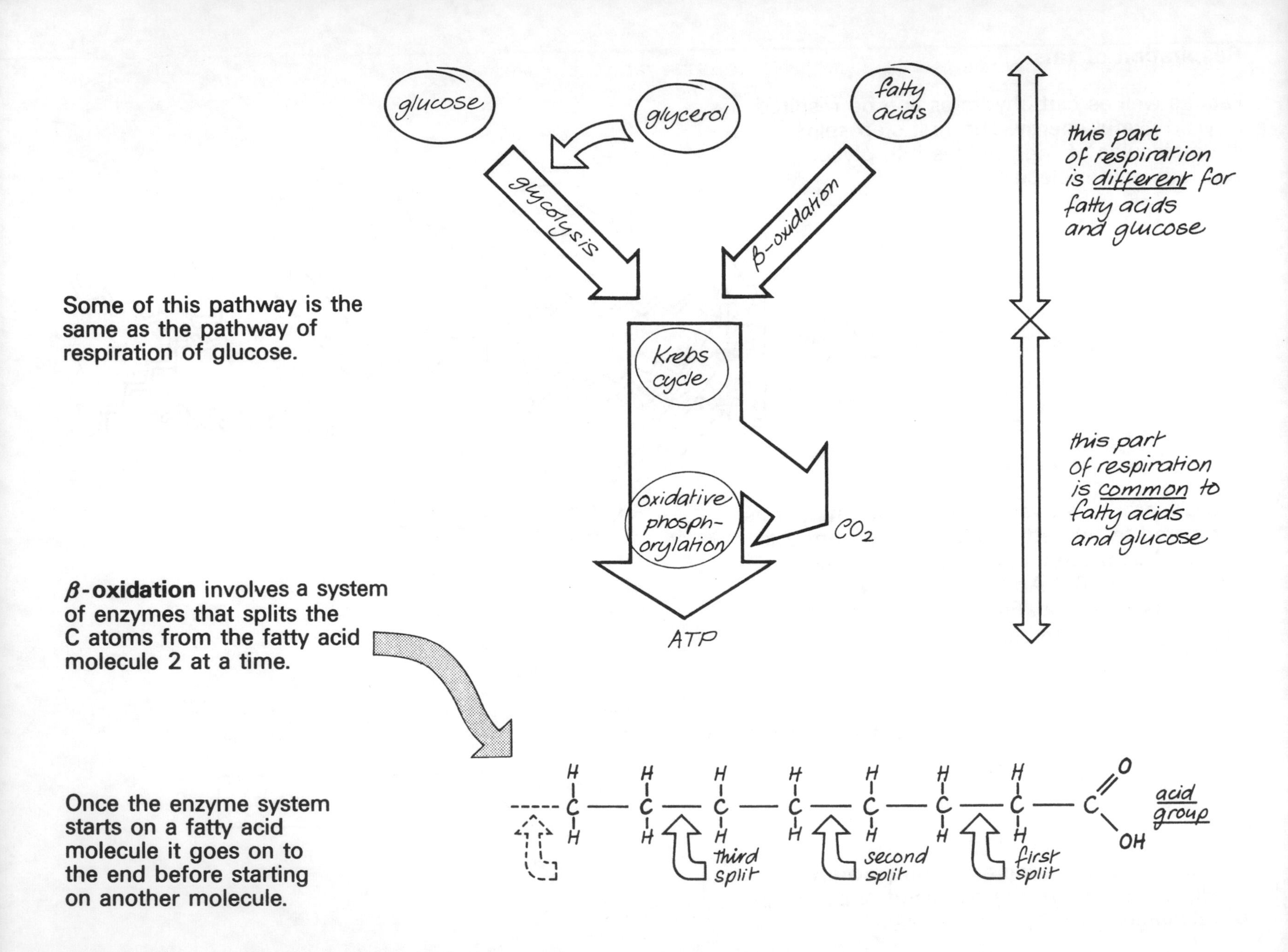
glucose
glycerol
fatty acids
glycolysis
β-oxidation
Krebs cycle
oxidative phosph-orylation
CO_2
ATP
this part of respiration is different for fatty acids and glucose
this part of respiration is common to fatty acids and glucose
H H H H H H H O
C C C C C C C
H H H H H H H OH
third split
second split
first split
acid group

Some of this pathway is the same as the pathway of respiration of glucose.

β-oxidation involves a system of enzymes that splits the C atoms from the fatty acid molecule 2 at a time.

Once the enzyme system starts on a fatty acid molecule it goes on to the end before starting on another molecule.

Each split is **oxidative**: enzymes pass the H from the fatty acid molecule (and also from some water molecules in the cell) to coenzyme molecules. One of the coenzymes is NAD, but another coenzyme with the symbol FAD takes part also:

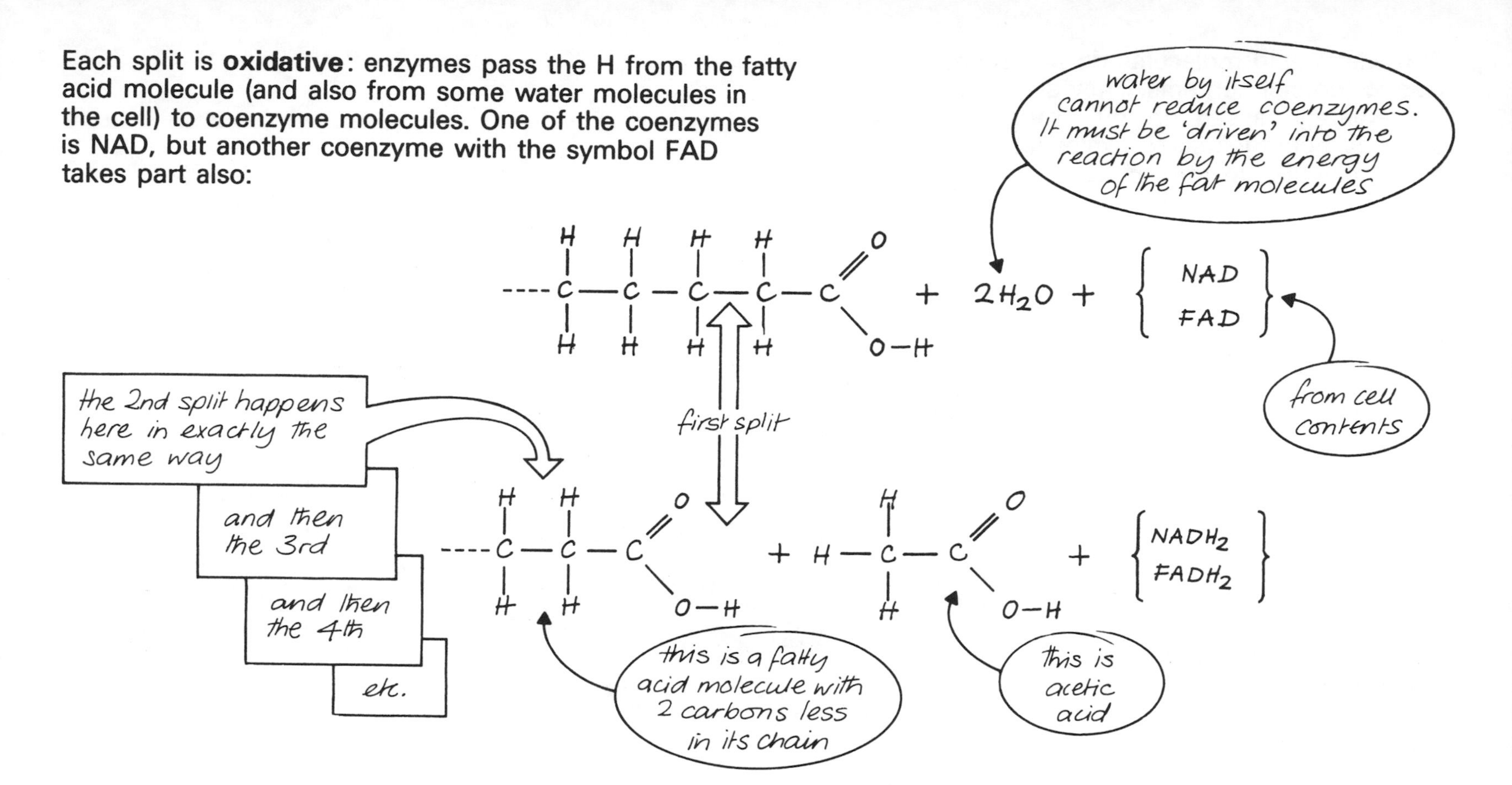

water by itself cannot reduce coenzymes. It must be 'driven' into the reaction by the energy of the fat molecules
H H H H O
----C—C—C—C—C
H H H H O—H
+ 2H₂O +
NAD
FAD
from cell contents
first split
the 2nd split happens here in exactly the same way
and then the 3rd
and then the 4th
etc.
H H O
----C—C—C
H H O—H
+ H—C—C
H O
H O—H
+
NADH₂
FADH₂
this is a fatty acid molecule with 2 carbons less in its chain
this is acetic acid

For **palmitic acid** (C_{16}) there are **7 splits**. Each one generates **one $NADH_2$ molecule and one $FADH_2$ molecule**.

1 palmitic acid molecule ⟶ 8 acetic acid molecules (C_2) + $7NADH_2$ + $7 FADH_2$

For stearic acid (C_{18}) there are **8 splits**

1 stearic acid molecule ⟶ 9 acetic acid molecules + $8NADH_2$ + $8FADH_2$

The acetic acid molecules do not appear free in the cell contents. They are immediately passed into the Krebs cycle by being joined to oxaloacetic acid molecules to form citric acid.

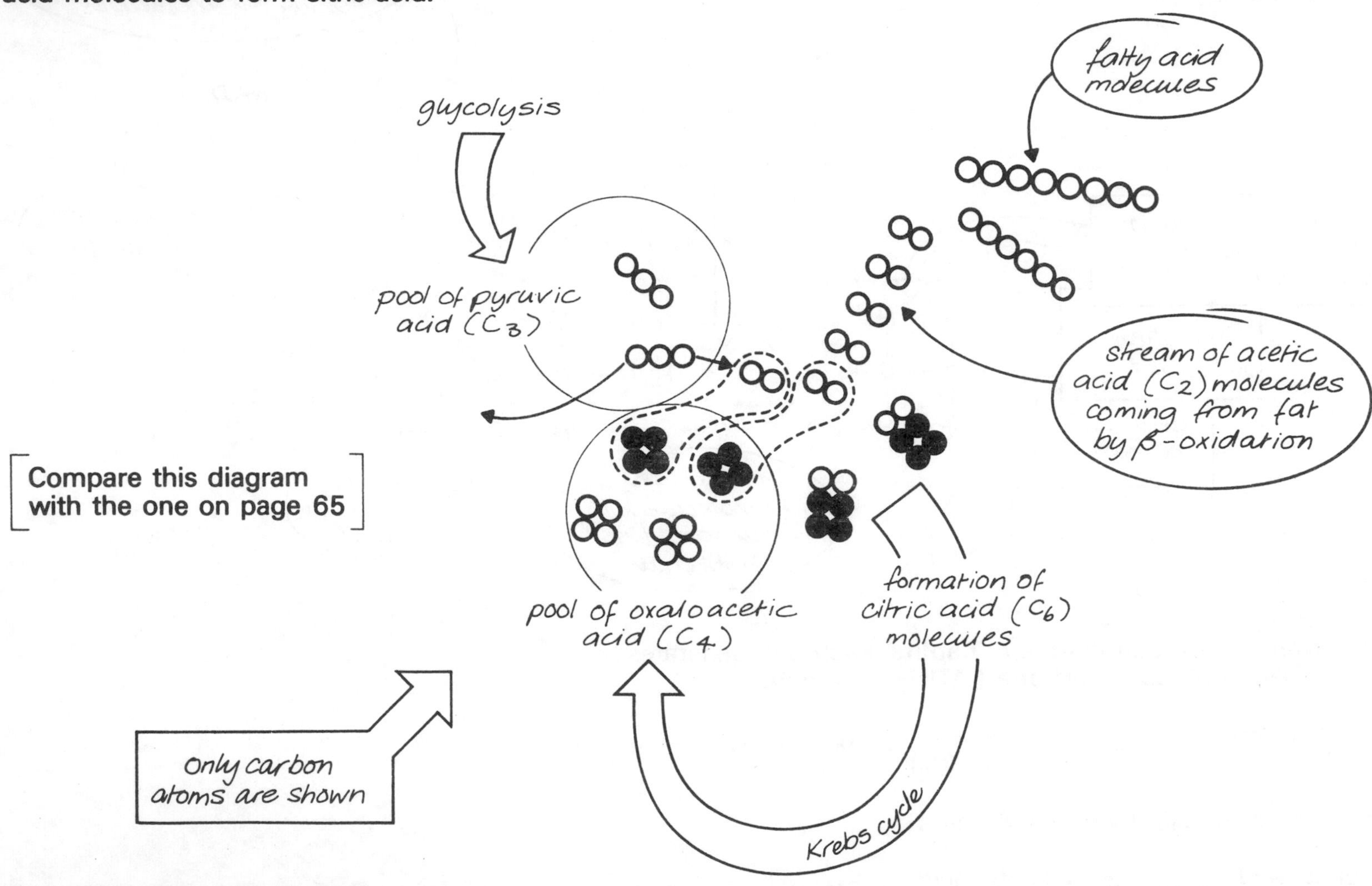
glycolysis
fatty acid molecules
pool of pyruvic acid (C_3)
stream of acetic acid (C_2) molecules coming from fat by β-oxidation
pool of oxaloacetic acid (C_4)
formation of citric acid (C_6) molecules
Krebs cycle
Only carbon atoms are shown

[Compare this diagram with the one on page 65]

The reduced coenzymes coming from β-oxidation can be oxidised to yield energy:

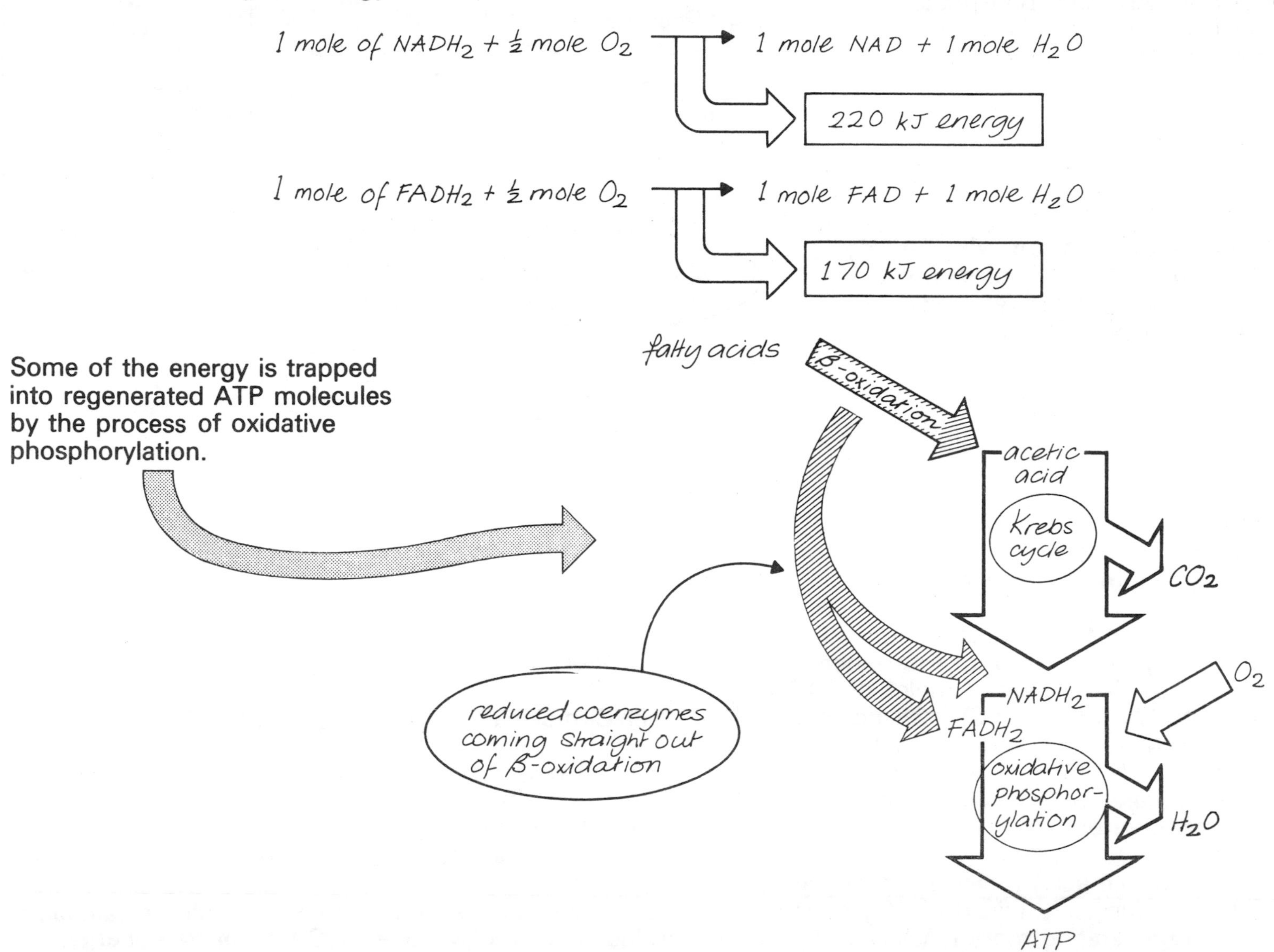
1 mole of $NADH_2$ + $\frac{1}{2}$ mole O_2
1 mole NAD + 1 mole H_2O
220 kJ energy
1 mole of $FADH_2$ + $\frac{1}{2}$ mole O_2
1 mole FAD + 1 mole H_2O
170 kJ energy
fatty acids
β-oxidation
acetic acid
Krebs cycle
CO_2
reduced coenzymes coming straight out of β-oxidation
$NADH_2$
$FADH_2$
O_2
oxidative phosphorylation
H_2O
ATP

Some of the energy is trapped into regenerated ATP molecules by the process of oxidative phosphorylation.

During oxidative phosphorylation one molecule of $NADH_2$ regenerates 3 ATP molecules from ADP and phosphate (see page 70).

3 ADP + 3 phosphate → 3 ATP

$NADH_2 + \frac{1}{2}O_2$ → NAD + H_2O

only about half of the energy is trapped with ATP (see page 60)

Oxidative phosphorylation with $FADH_2$ regenerates only 2 ATP molecules (**$FADH_2$** is the flavoprotein referred to as 'FPH_2' on pages 71 and 72: see glossary).

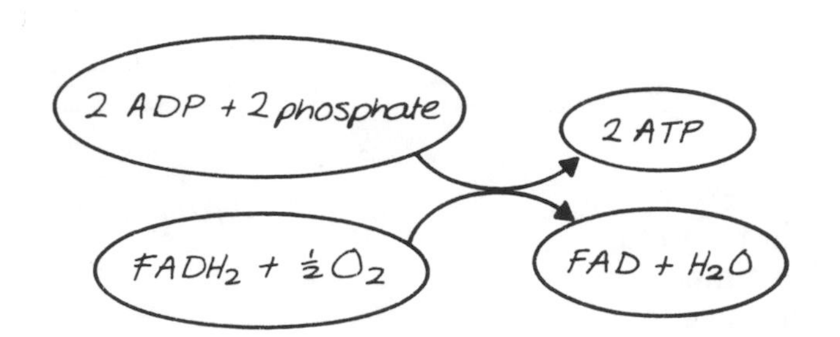

2 ADP + 2 phosphate
2 ATP
$FADH_2 + \frac{1}{2}O_2$
FAD + H_2O

Useful energy yield of β-oxidation

Palmitic acid, if oxidised completely to water and carbon dioxide, yields 10,000 kJ **total** energy per mole. 7 moles of $NADH_2$ and 7 moles of $FADH_2$ can regenerate $(7 \times 3) + (7 \times 2) = 35$ moles of ATP. 1 mole of ATP represents about 30 kJ of useful (free) energy. $\therefore$ 35 moles of ATP represents about 1000 kJ of free energy which is about $\frac{1000}{10,000}$ = **only 10% of the total energy**.

But – the C_2 (acetic acid) molecules produced by β-oxidation are further oxidised to CO_2 by the Krebs cycle, regenerating more $NADH_2$ in the process. This $NADH_2$ also regenerates ATP by oxidative phosphorylation.

Each C_2 molecule is oxidised, generating $4NADH_2$ molecules.

So – for one molecule of palmitic acid (producing 8 C_2 molecules), $(8 \times 4) = 32$ molecules of $NADH_2$ are formed by the Krebs cycle.
This in turn gives 32×3 = about 100 ATP molecules.

So this part of the oxidation of one mole of palmitic acid yields about $100 \times 30 = 3,000$ kJ of free energy.

The **total yield** of useful energy (in the form of ATP) is therefore

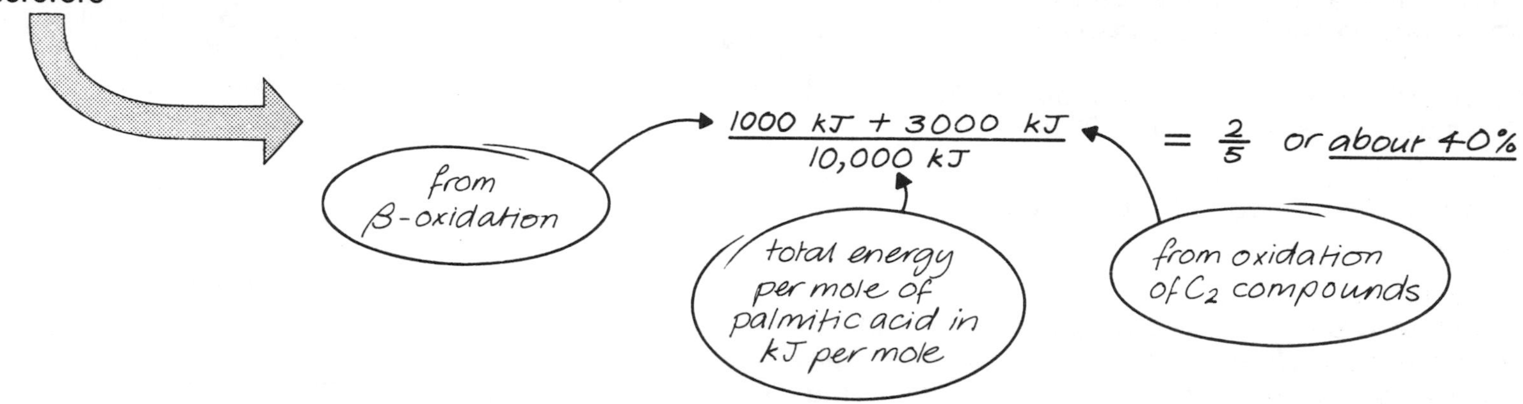

1000 kJ + 3000 kJ
10,000 kJ
= 2/5 or about 40%
from β-oxidation
total energy per mole of palmitic acid in kJ per mole
from oxidation of C_2 compounds

This is very similar to the efficiency of glucose respiration

Note Although the efficiency of conversion of energy from fat or carbohydrate molecules to energy in ATP molecules is about the same (40%) the **actual amount** of energy is rather different. This is because fats contain more energy than carbohydrates:

The total possible free energy per mole of glucose is 2900 kJ
" " " " " for each C atom (per mole) is $\frac{2900}{6}$ = 483 kJ
40% of this is 193 kJ
" " " " " per mole of palmitic acid is 10,000 kJ
" " " " " for each C atom (per mole) is $\frac{10,000}{16}$ = 625 kJ
40% of this is 250 kJ

This is because fats are **more reduced** than carbohydrates

The **more reduced** a carbon compound is, the **more** useful energy it contains.
The **more oxidised** a carbon compound is, the **less** useful energy it contains.
CO_2 (the most oxidised form of carbon) contains no useful energy at all.

Fats are a more compact way of storing energy than carbohydrates are. (Fats are commonly found as storage food in seeds and eggs.)

1 kg of carbohydrate contains about 33 'moles' of $[CH_2O]$ with a total free energy of 16,000 kJ.

1 kg of fat contains about 71 'moles' of $[CH_2]$ with a total free energy of 44,500 kJ – nearly 3 times as much.

assuming this to be the 'average' formula for a carbohydrate

assuming this to be the 'average' formula for a fatty acid

Fat synthesis
The fatty acid chain grows two carbon atoms at a time, but not merely by reversal of β-oxidation. The process is, of course, **reductive**, and the coenzyme $NADPH_2$ (**not** $NADH_2$ or $FADH_2$) is used to provide the reducing power ($NADPH_2$ can be generated by the pentose phosphate pathway – PPP – of respiration).

The building units are not acetic acid (C_2) groups, but surprisingly malonic acid (C_3) groups.

H
H—C—H
—C=O

acetyl group (from acetic acid)

COOH
H—C—H
—C=O

malonyl group (from malonic acid)

But the malonyl groups are formed from acetyl groups + CO_2

acetyl group + CO_2 + ATP ⟶ malonyl group + ADP + phosphate.

When the fatty acid chain is made, the CO_2 molecule which was used to make the malonyl group from the acetyl group, is split off again as gaseous CO_2.

$$----CH_2-CH_2-CO- \;+\; \begin{matrix}COOH\\ |\\ CH_2\\ |\\ -C=O\end{matrix} \longrightarrow -----CH_2-CH_2-CH_2-CH_2-CO-$$

this is a fatty acid molecule being formed

oxygen extracted by $NADPH_2$ (to give NADP + H_2O)

this is a reduction

split off as CO_2

carbon chain lengthened by 2 atoms

Why is CO_2 involved?

Or why are acetyl groups not used directly in fatty acid synthesis?

Because:

the energy relationships of the equation allow it to go rapidly to the right (it is exergonic). If acetyl groups are substituted for malonyl groups the energy relationships are different, and the reaction will not 'go'. Note that energy from ATP is used to make malonyl from acetyl, so the 'equation' is really 'driven' by energy from ATP.

The acetyl groups come from glycolysis.

If excess glucose is present (more than is needed for provision of respiratory energy), some of it will be channelled into fat synthesis – so over-consumption of carbohydrate helps make people fat!

The final step in fat synthesis links fatty acids to glycerol; which is also derived from glycolysis (from triose phosphate).

3 fatty acid molecules + 1 glycerol molecule ⟶ fat molecule

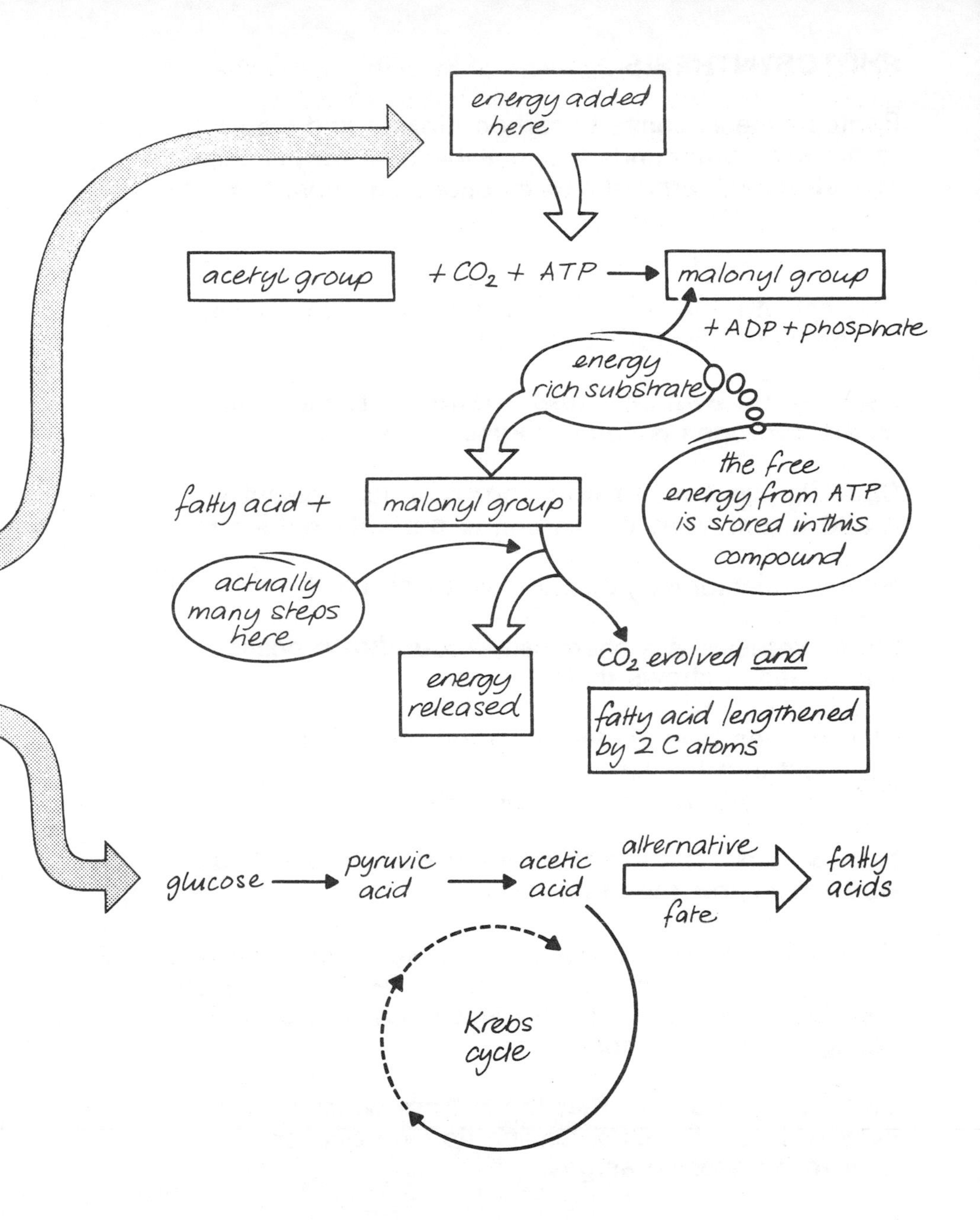
energy added here
acetyl group
+ CO_2 + ATP ⟶
malonyl group
+ ADP + phosphate
energy rich substrate
the free energy from ATP is stored in this compound
fatty acid +
malonyl group
actually many steps here
energy released
CO_2 evolved and
fatty acid lengthened by 2 C atoms
glucose
pyruvic acid
acetic acid
alternative
fate
fatty acids
Krebs cycle

PHOTOSYNTHESIS

Photosynthesis converts carbon dioxide and water to organic compounds – sugar (hexose) for instance. The **stoichiometry** of this reaction can be written as

CO_2 has **no available free energy** in its molecules as the carbon is **fully oxidised**.

H_2O has **no available free energy** in its molecules as the hydrogen is **fully oxidised**.

$C_6H_{12}O_6$ has **a lot of free energy** in its molecules (2900 kJ/mole) as the carbon is **partially reduced**.

So the stoichiometry of photosynthesis can be written

Photosynthesis is a chemically **reductive** process. The equation shows that:

1. oxygen is extracted from the reactants and lost into the air,
2. the carbon has hydrogen attached to it.

But as usual, the stoichiometric equation tells us nothing of the mechanism.

Research has shown that all the oxygen gas comes from water. None comes from the CO_2. So the first step in considering the mechanism leads to a change in the equation:

This equation shows that the H from water chemically reduces the CO_2, leaving its own oxygen over to be evolved as gas.

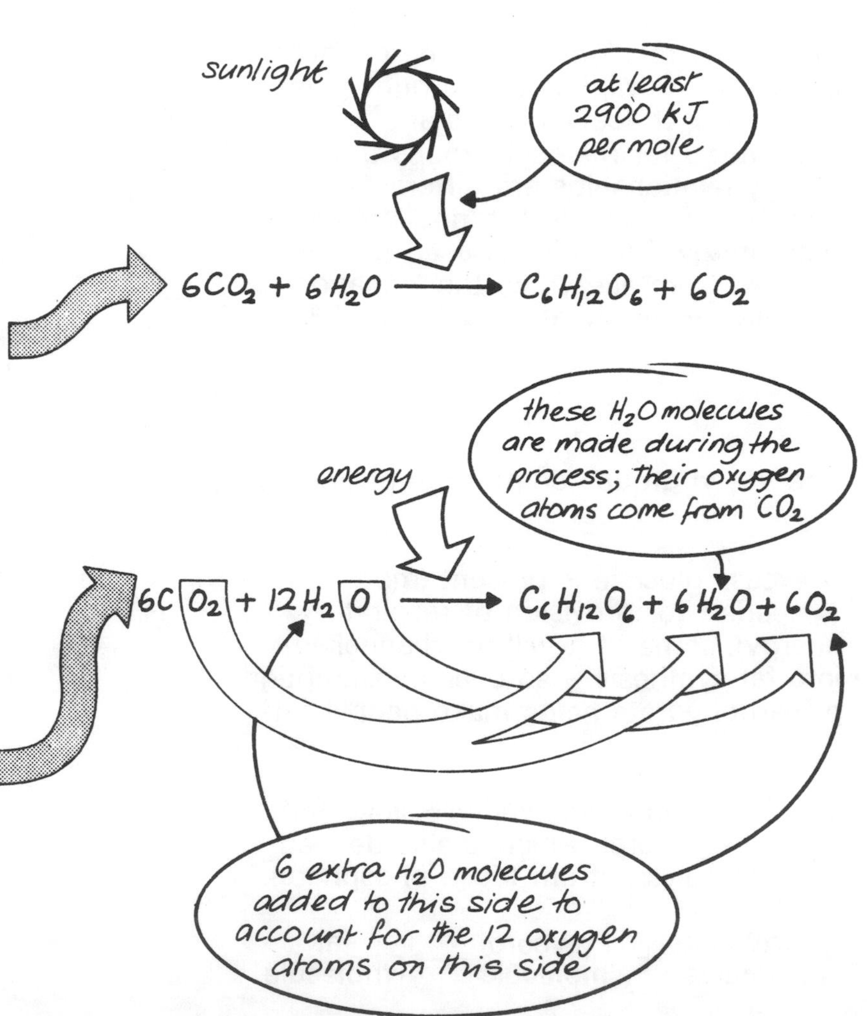
6 moles of CO_2 + 6 moles of H_2O → 1 mole $C_6H_{12}O_6$ + 6 moles of O_2
sunlight
at least 2900 kJ per mole
$6CO_2 + 6H_2O \rightarrow C_6H_{12}O_6 + 6O_2$
energy
these H_2O molecules are made during the process; their oxygen atoms come from CO_2
$6CO_2 + 12H_2O \rightarrow C_6H_{12}O_6 + 6H_2O + 6O_2$
6 extra H_2O molecules added to this side to account for the 12 oxygen atoms on this side

It is easier to see if the equation is divided by 6:

energy

this symbol means 1/6th of a hexose molecule

$$CO_2 + 2H_2O^* \longrightarrow [CH_2O] + H_2O + O_2^*$$

this water is used in the process

this water is generated by the process

This equation shows that each CO_2 molecule needs 4 H atoms to reduce it. The H's come from water.

from water

$$CO_2 + 4[\text{H atoms}] \longrightarrow [CH_2O] + H_2O$$

two of these extract an oxygen from CO_2

the other two combine with what's left

both of these are chemical reductions of CO_2

To provide the H atoms, water must be split.

It is the splitting of water molecules to give H atoms which requires a lot of energy. The H atoms carry this energy with them, and some of it finishes up locked into the sugar molecules.

Note: The splitting of water in photosynthesis must not be confused with the ionisation of water.

Conceptual reminder

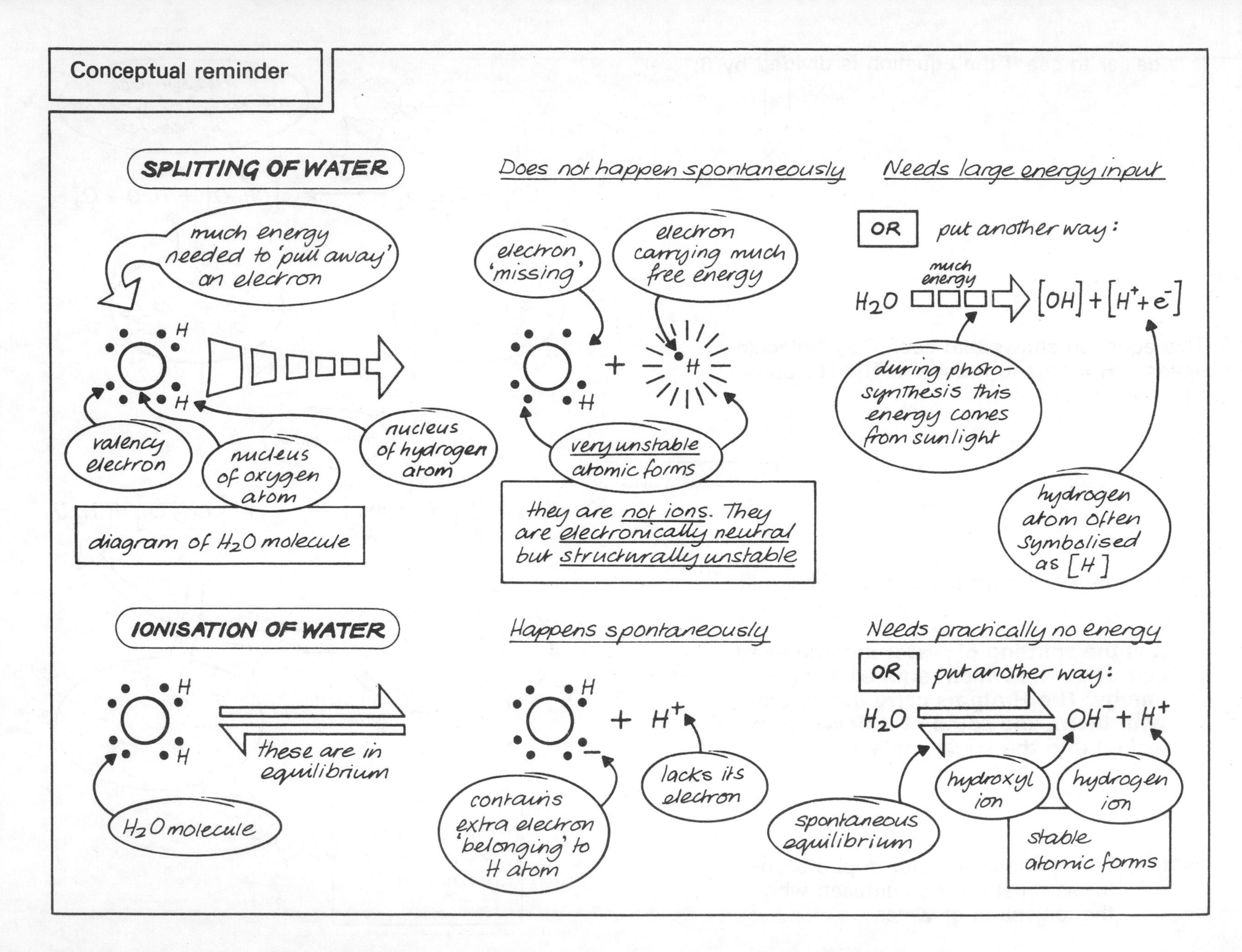
SPLITTING OF WATER
Does not happen spontaneously
Needs large energy input
much energy needed to 'pull away' an electron
H
H
valency electron
nucleus of oxygen atom
nucleus of hydrogen atom
diagram of H_2O molecule
electron 'missing'
electron carrying much free energy
H
H
very unstable atomic forms
they are not ions. They are electronically neutral but structurally unstable
OR
put another way:
much energy
H_2O → [OH] + [H^+ + e^-]
during photosynthesis this energy comes from sunlight
hydrogen atom often symbolised as [H]
IONISATION OF WATER
Happens spontaneously
Needs practically no energy
H
H
these are in equilibrium
H_2O molecule
H
+ H^+
contains extra electron 'belonging' to H atom
lacks its electron
OR
put another way:
H_2O ⇌ OH^- + H^+
spontaneous equilibrium
hydroxyl ion
hydrogen ion
stable atomic forms

Neither the [OH]'s nor the [H]'s formed from the splitting of water really exist.

The [OH]'s instantly combine together in groups, 4 at a time, like this.

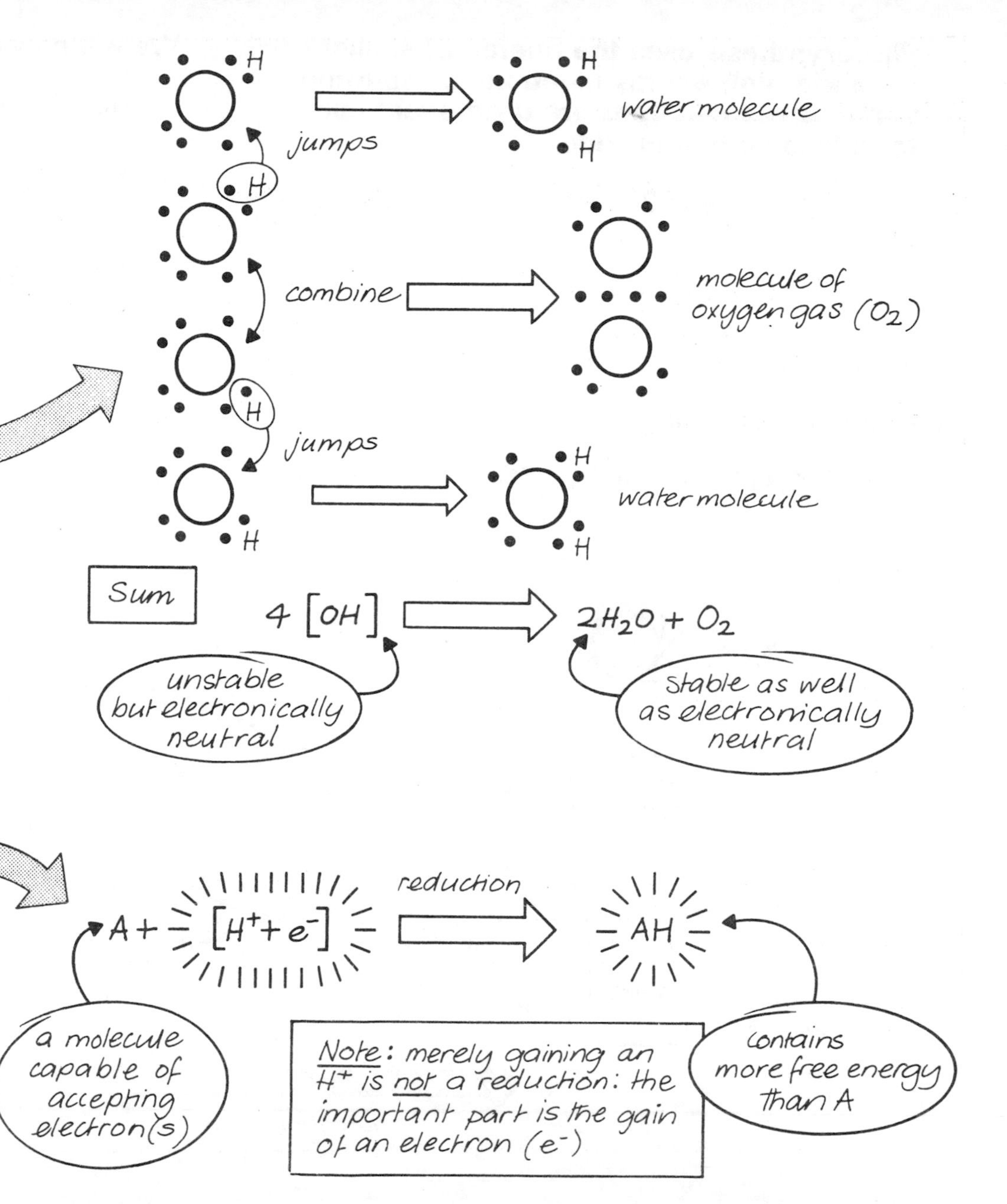

H
jumps
H
combine
H
jumps
H
water molecule
molecule of
oxygen gas (O_2)
water molecule
Sum
4 [OH]
$2H_2O + O_2$
unstable
but electronically
neutral
stable as well
as electronically
neutral
reduction
A + [H^++e^-]
AH
a molecule
capable of
accepting
electron(s)
Note: merely gaining an
H^+ is not a reduction: the
important part is the gain
of an electron (e^-)
contains
more free energy
than A

The [H^+ + e^-]'s find molecules of another compound to combine with: **because this compound gains an electron** (as well as an H^+) **it becomes chemically reduced:**

Photosynthesis uses the energy of sunlight to generate high energy electrons by splitting water: these electrons are used eventually to reduce carbon dioxide.

The equation on page 85 can now be written again because:

$$4H_2O \xrightarrow{\text{sunlight}} 4[H] + 4[OH] \Rightarrow 2H_2O + O_2$$

$$\text{then } CO_2 + 4[H] \longrightarrow [CH_2O] + H_2O$$

The sum of these being:

$$\mathbf{CO_2 + 4H_2O \xrightarrow{\text{sunlight}} [CH_2O] + 3H_2O + O_2}$$

We will now consider

How is sunlight being used **?**

How is water split **?**

How is CO_2 biochemically reduced to form sugar **?**

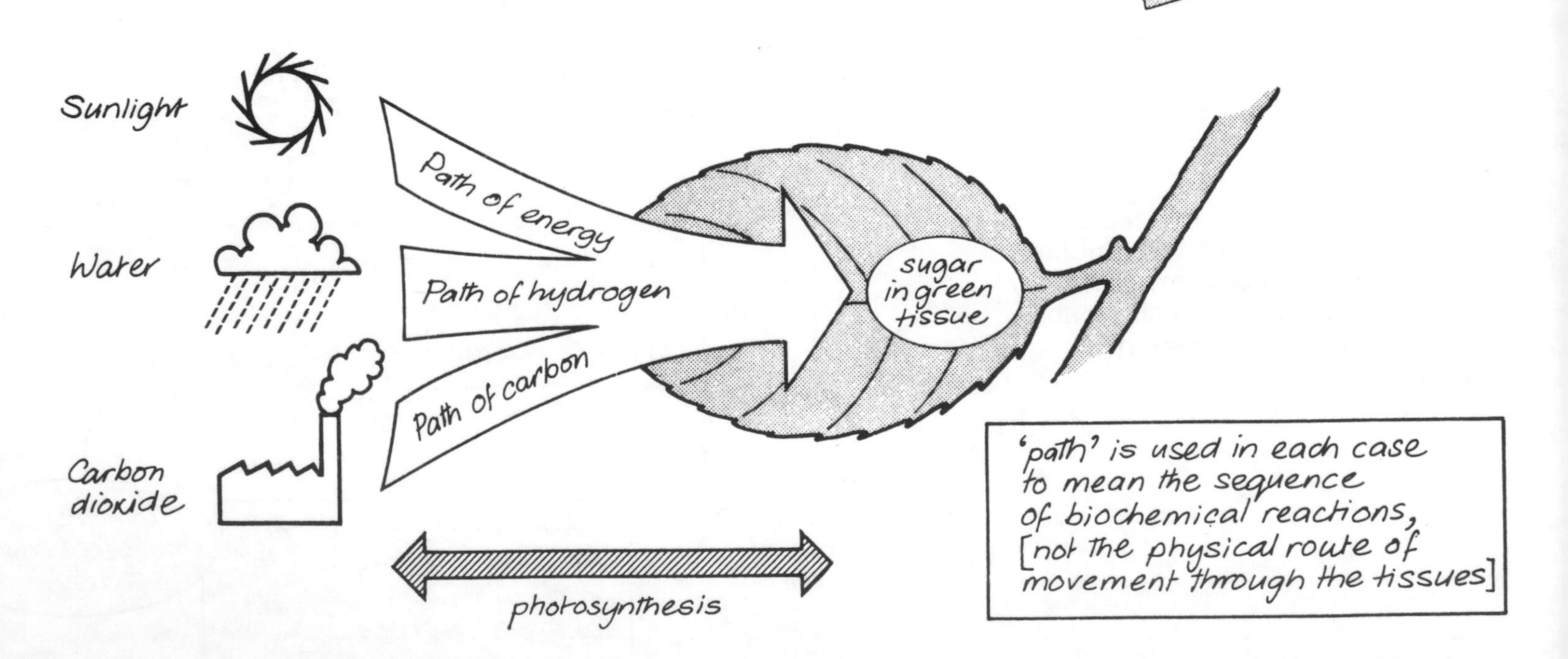
OR
Sunlight
Water
Carbon dioxide
Path of energy
Path of hydrogen
Path of carbon
sugar in green tissue
photosynthesis
'path' is used in each case to mean the sequence of biochemical reactions, [not the physical route of movement through the tissues]

A conceptual diversion about sunlight

① Sunlight is a form of energy. All light is a form of energy.

② Light exists both as waves and as particles **at the same time.**
It is impossible to understand this by comparison with anything we experience with our senses. We must just accept it from the physicists!

③ The **waves/particles** of light are called **photons.** Photons are the smallest possible particles of light. They are therefore **indivisible** (if they were divisible there would be smaller particles!). They can be considered to be 'atoms' of light.

④ **Photons have wavelengths**. There are different 'types' of photons (just as there are different types of atoms). Their difference lies in the **amount of energy** they consist of, and also in their **wavelength**.

Photons with short wavelengths consist of more energy than photons with long wavelengths.

A photon with **half** the wavelength of another photon, consists of **double** the energy. One with $\frac{1}{3}$rd the wavelength, consists of 3 times the energy, and so on.

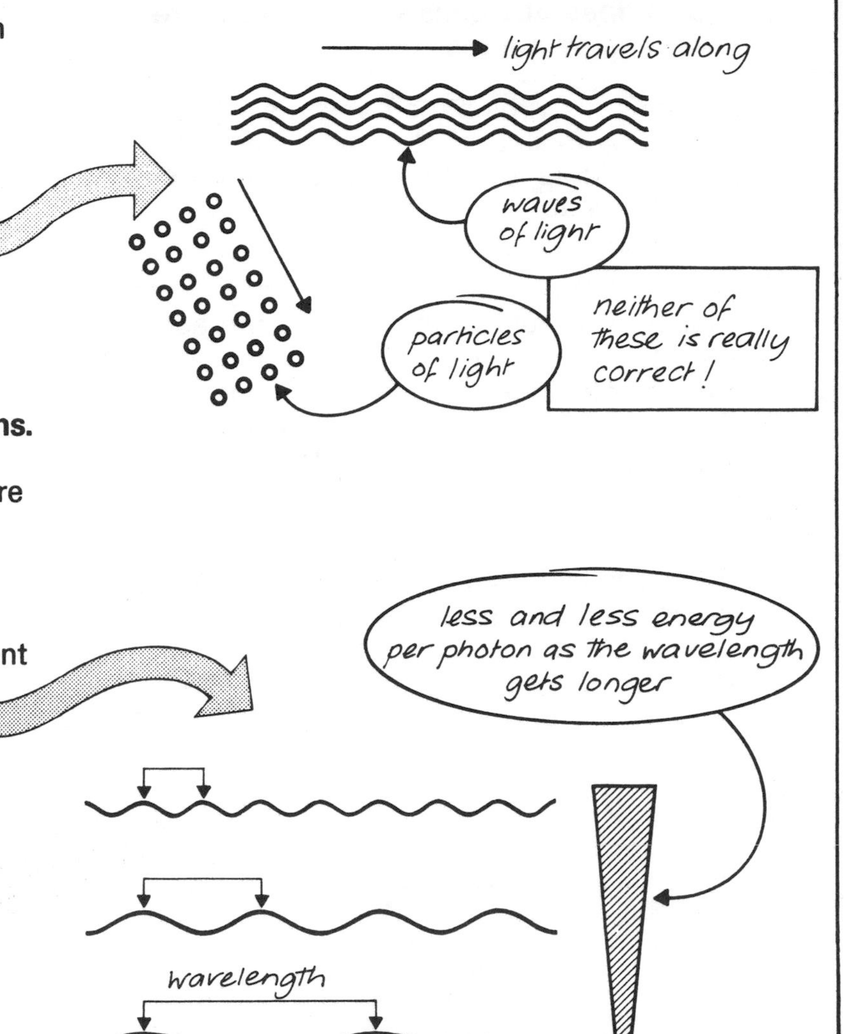

light travels along
waves of light
particles of light
neither of these is really correct!
less and less energy per photon as the wavelength gets longer
wavelength

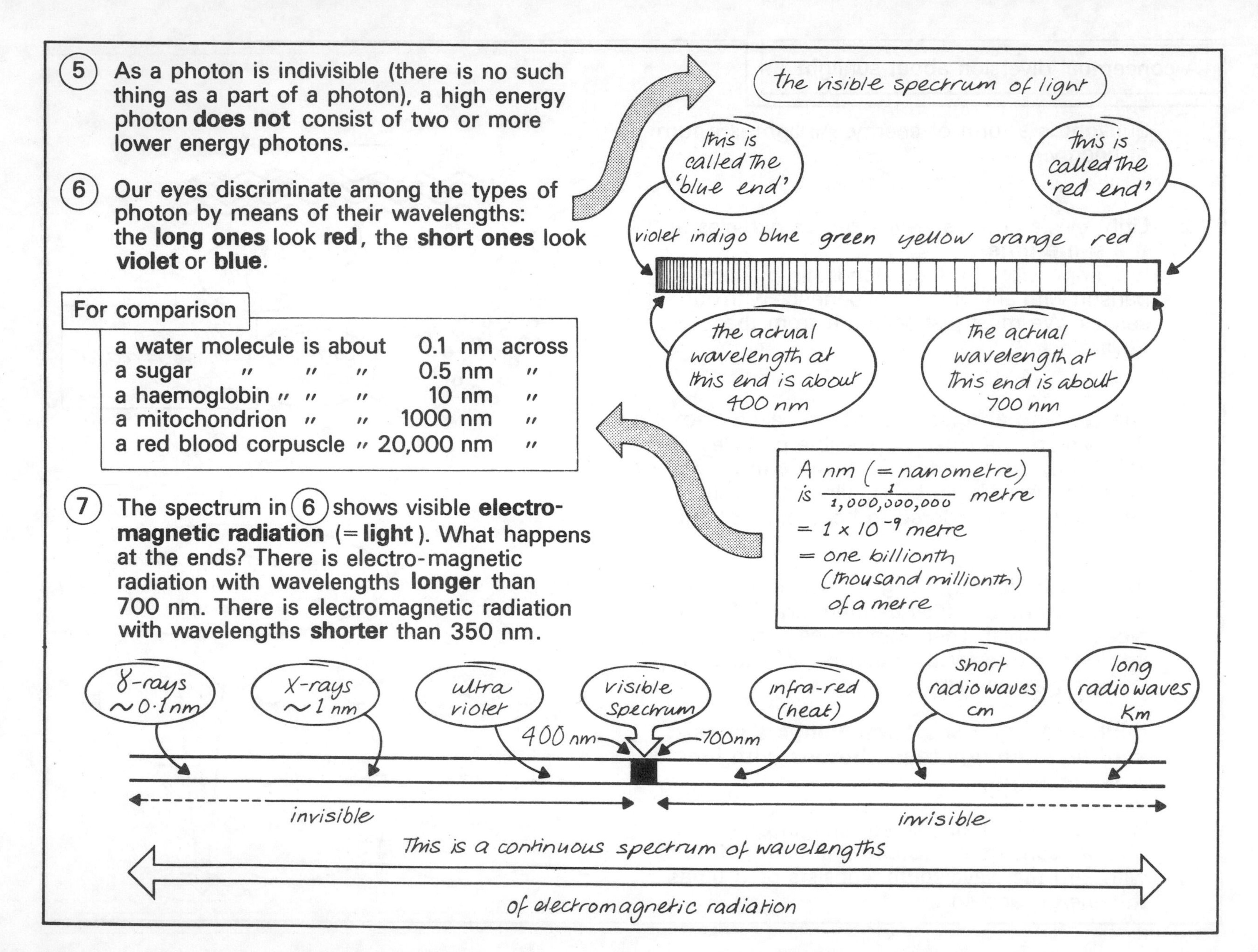
5 As a photon is indivisible (there is no such thing as a part of a photon), a high energy photon does not consist of two or more lower energy photons.
6 Our eyes discriminate among the types of photon by means of their wavelengths: the long ones look red, the short ones look violet or blue.
For comparison
a water molecule is about 0.1 nm across
a sugar " " " 0.5 nm "
a haemoglobin " " " 10 nm "
a mitochondrion " " 1000 nm "
a red blood corpuscle " 20,000 nm "
7 The spectrum in 6 shows visible electro-magnetic radiation (= light). What happens at the ends? There is electro-magnetic radiation with wavelengths longer than 700 nm. There is electromagnetic radiation with wavelengths shorter than 350 nm.
the visible spectrum of light
this is called the 'blue end'
this is called the 'red end'
violet indigo blue green yellow orange red
the actual wavelength at this end is about 400 nm
the actual wavelength at this end is about 700 nm
A nm (= nanometre) is 1/1,000,000,000 metre
= 1 x 10⁻⁹ metre
= one billionth (thousand millionth) of a metre
γ-rays ~0·1 nm
X-rays ~1 nm
ultra violet
visible spectrum
infra-red (heat)
short radio waves cm
long radio waves Km
400 nm
700nm
invisible
invisible
This is a continuous spectrum of wavelengths
of electromagnetic radiation

The 'particles' of electromagnetic radiation have the general name **quanta**, singular **quantum**. **Photons are merely visible quanta**. The energy content of a quantum of radiowave (very long wavelength) is very low. The energy content of a quantum of X-ray (very short wavelength) is enormously higher.

Example

a quantum of X-rays of wavelength 1 nm has one million million (10^{12}) × as much energy as a quantum of radiowave of wavelength 1 km.

(8) For the energy of a quantum to be used it must be absorbed.

Transparent objects do not absorb photons. Any chemical compound which does absorb photons is called a pigment.

Pigments which absorb all visible wavelengths **must be black** (there is no light left over for the eye to see). Other pigments absorb only part of the visible spectrum–

the colour we see is the light which is **not** absorbed.

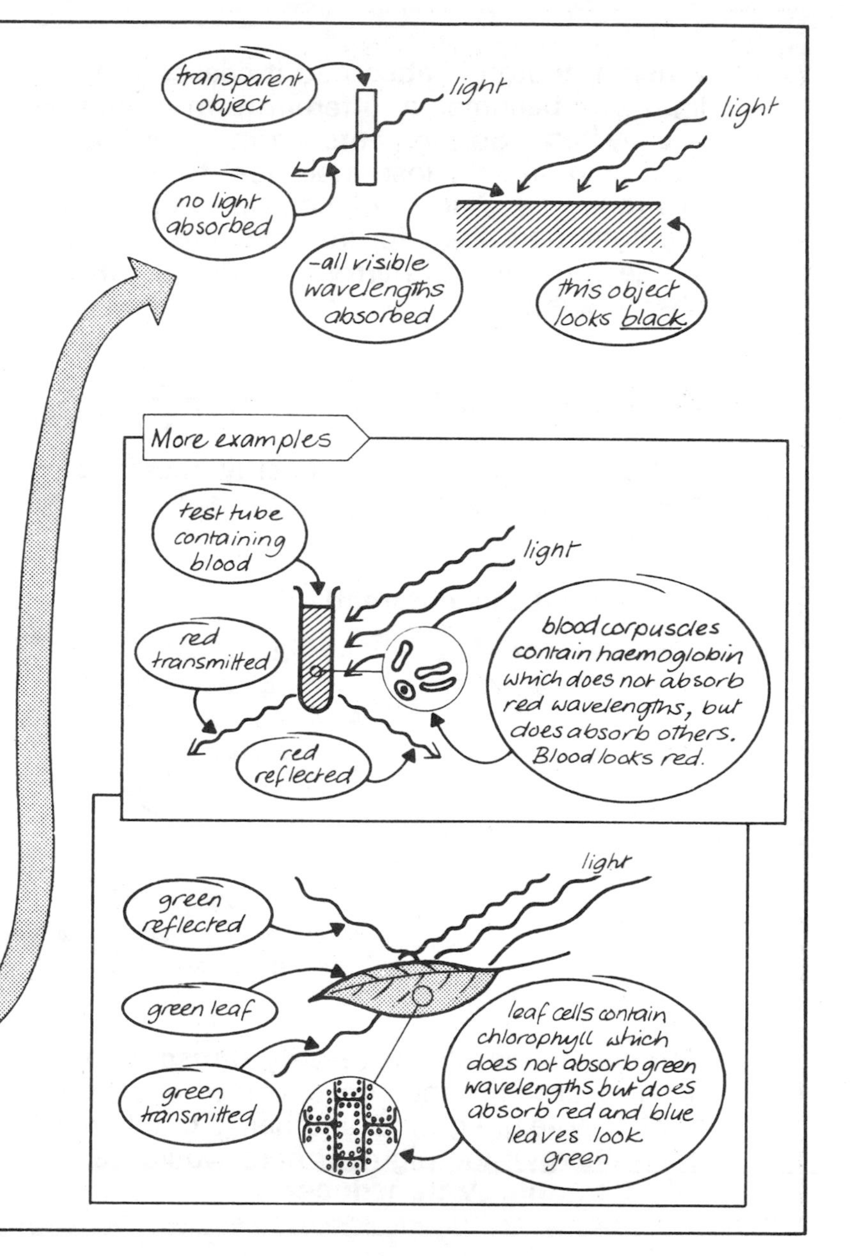
transparent object
light
no light absorbed
light
-all visible wavelengths absorbed
this object looks black
More examples
test tube containing blood
light
red transmitted
blood corpuscles contain haemoglobin which does not absorb red wavelengths, but does absorb others. Blood looks red.
red reflected
light
green reflected
green leaf
leaf cells contain chlorophyll which does not absorb green wavelengths but does absorb red and blue leaves look green
green transmitted

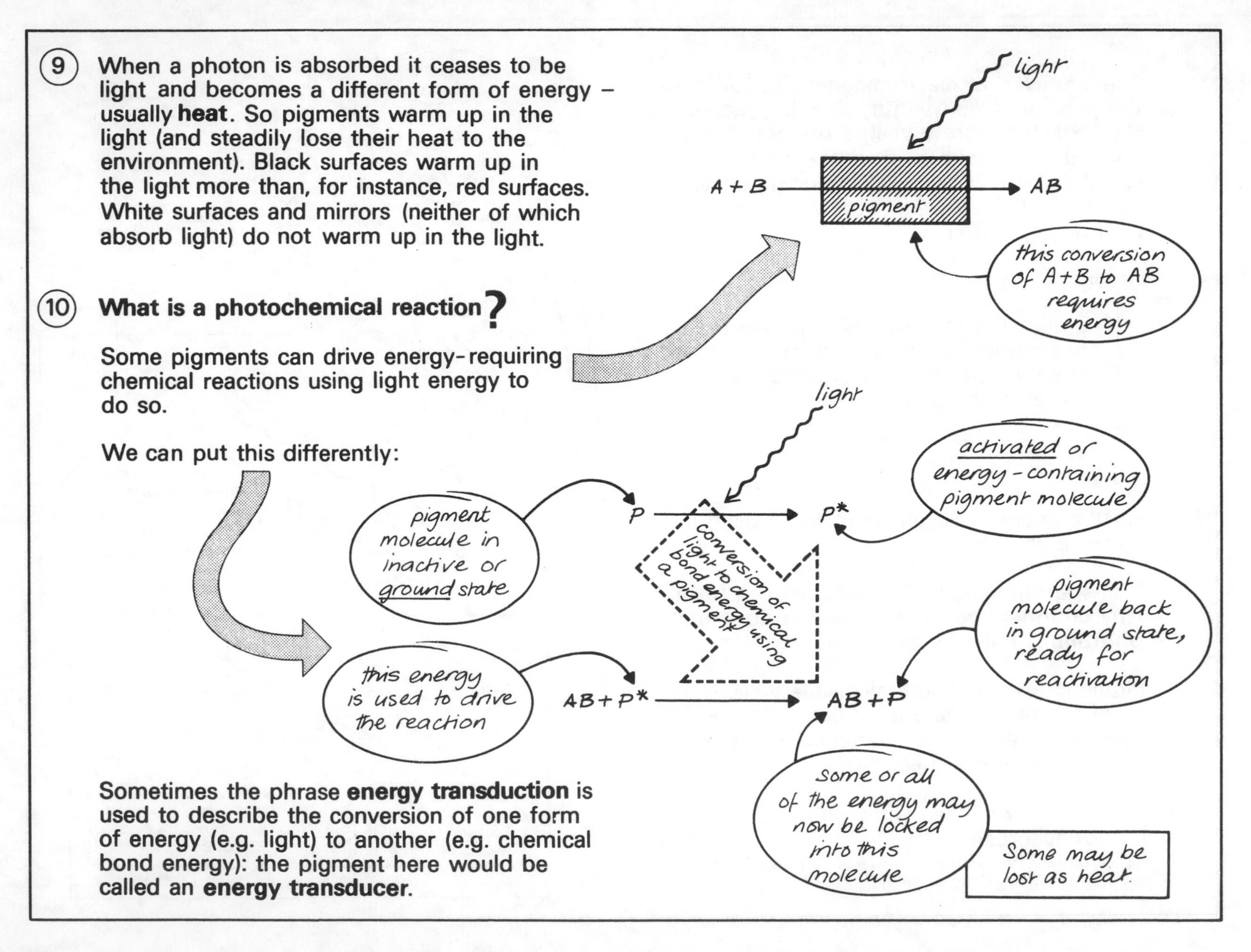
light
A + B
pigment
AB
this conversion of A+B to AB requires energy
light
pigment molecule in inactive or ground state
P
P*
activated or energy-containing pigment molecule
conversion of light to chemical bond energy using a pigment
this energy is used to drive the reaction
AB+P*
AB+P
pigment molecule back in ground state, ready for reactivation
some or all of the energy may now be locked into this molecule
Some may be lost as heat.

(9) When a photon is absorbed it ceases to be light and becomes a different form of energy – usually **heat**. So pigments warm up in the light (and steadily lose their heat to the environment). Black surfaces warm up in the light more than, for instance, red surfaces. White surfaces and mirrors (neither of which absorb light) do not warm up in the light.

(10) **What is a photochemical reaction?**

Some pigments can drive energy-requiring chemical reactions using light energy to do so.

We can put this differently:

Sometimes the phrase **energy transduction** is used to describe the conversion of one form of energy (e.g. light) to another (e.g. chemical bond energy): the pigment here would be called an **energy transducer**.

(11) How much energy is there in light?

A pigment molecule will absorb photons one at a time. [It cannot absorb part of a photon as photons are indivisible.]
When we consider the **energy requirement** of the reaction,

$$A + B \longrightarrow AB$$

we talk about **kilojoules per mole** and **not** about the energy required by single molecules. This is for convenience – the energy relationships (and weight, and everything else!) of single molecules are too tiny to measure or even to think about. We scale everything up for 602,000,000,000,000,000,000,000 molecules
= 6.02×10^{23} [this was chosen as it is the number of molecules in]

= 1 gram molecule (mole) of any compound

1 mole (18g) of water contains 6.02×10^{23} molecules
1 mole (44g) of carbon dioxide contains 6.02×10^{23} molecules
1 mole (180g) of hexose contains 6.02×10^{23} molecules.

So if $P \longrightarrow P^*$ requires one photon we can say 1 mole of $P \longrightarrow$ 1 mole of P^* requires

6.02×10^{23} photons = 1 'mole-photon'.

The energy of a mole-photon (or more generally a **mole-quantum**) is called an **einstein:** its actual value varies with the wavelength.

An einstein of red light with wavelength 700 nm consists of 170 kJ of energy
An einstein of ultraviolet with wavelength 350 nm consists of 340 kJ of energy
An einstein of radiowave with wavelength 7 cm consists of 0.017 kJ of energy
An einstein of X-ray with wavelength 7 nm consists of 17,000 kJ of energy.

Note If a mole each of A and B are converted to AB by mediation of a pigment, **a mole of pigment is not required**. As long as pigment molecules are converted from the ground state to the activated state sufficiently often to allow a mole quantum of light to be chanelled into the reaction, the reaction is able to proceed. The pigment acts **catalytically**.

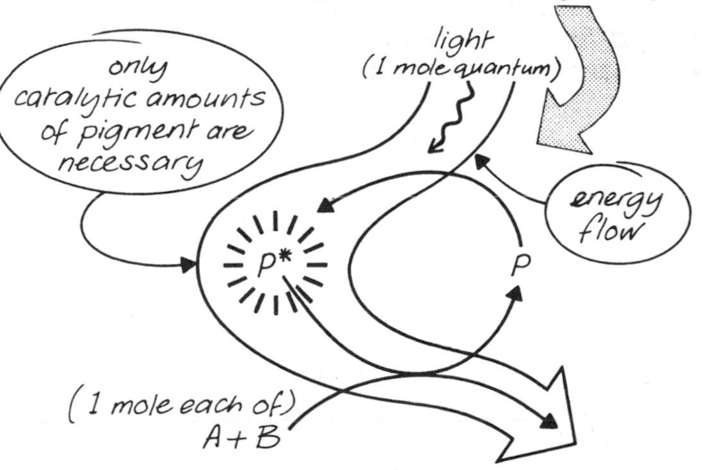

only catalytic amounts of pigment are necessary
light (1 mole quantum)
energy flow
P*
P
(1 mole each of) A+B

NOW FOR

The path of energy in photosynthesis

It is convenient to deal with this together with the path of [H] as energy is frequently transferred by means of high-energy electrons (remember that [H] is the same as $[H^+ + e^-]$).

Chlorophyll molecules absorb light at the red (long wavelength) and blue (short wavelength) ends of the spectrum – that is why chlorophyll looks green. When we measure the amount of energy absorbed by a chlorophyll solution at various wavelengths (by using variously coloured light), we can draw a graph called **the absorption spectrum**.

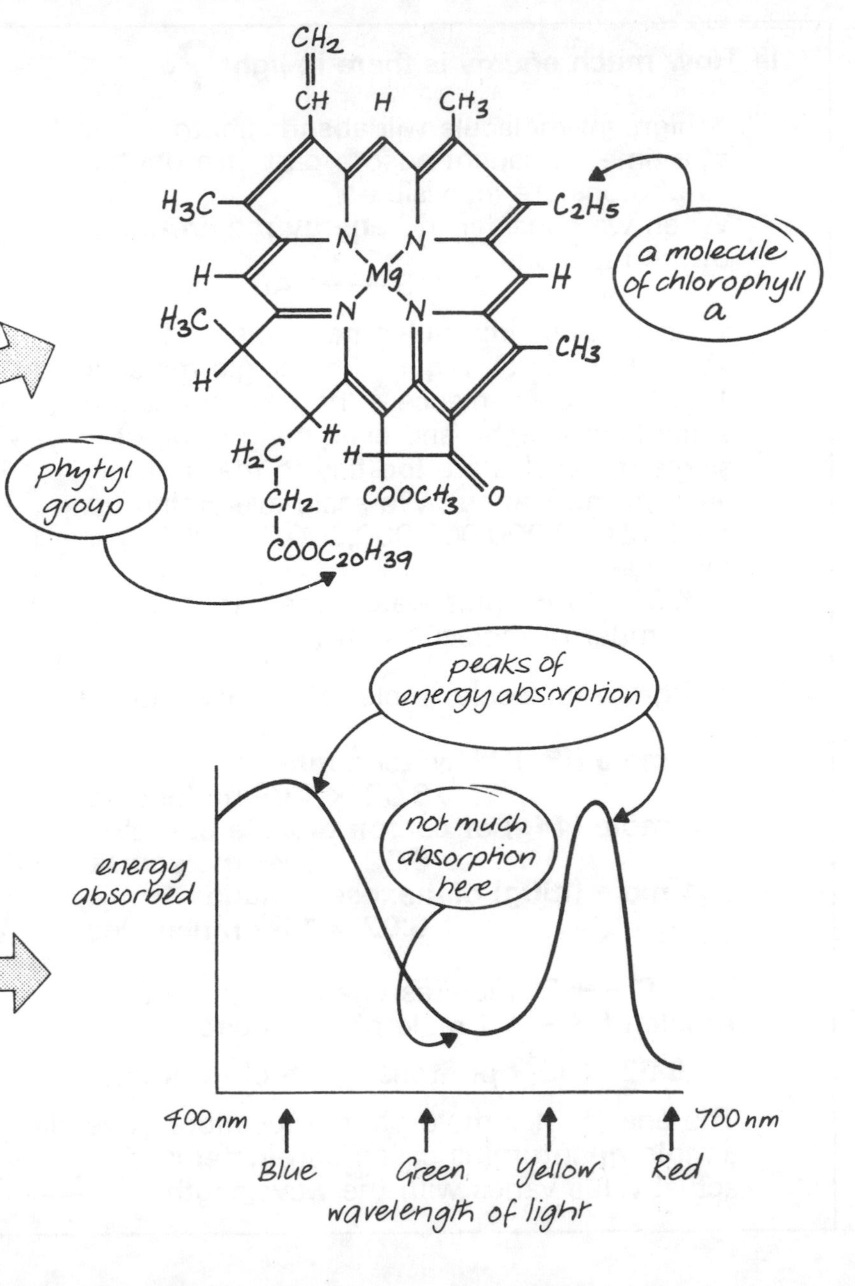

a molecule of chlorophyll a
phytyl group
peaks of energy absorption
not much absorption here
energy absorbed
400 nm
700 nm
Blue
Green
Yellow
Red
wavelength of light

When we measure the amount of photosynthesis which goes on in a leaf at the various wavelengths we get an **action spectrum**.

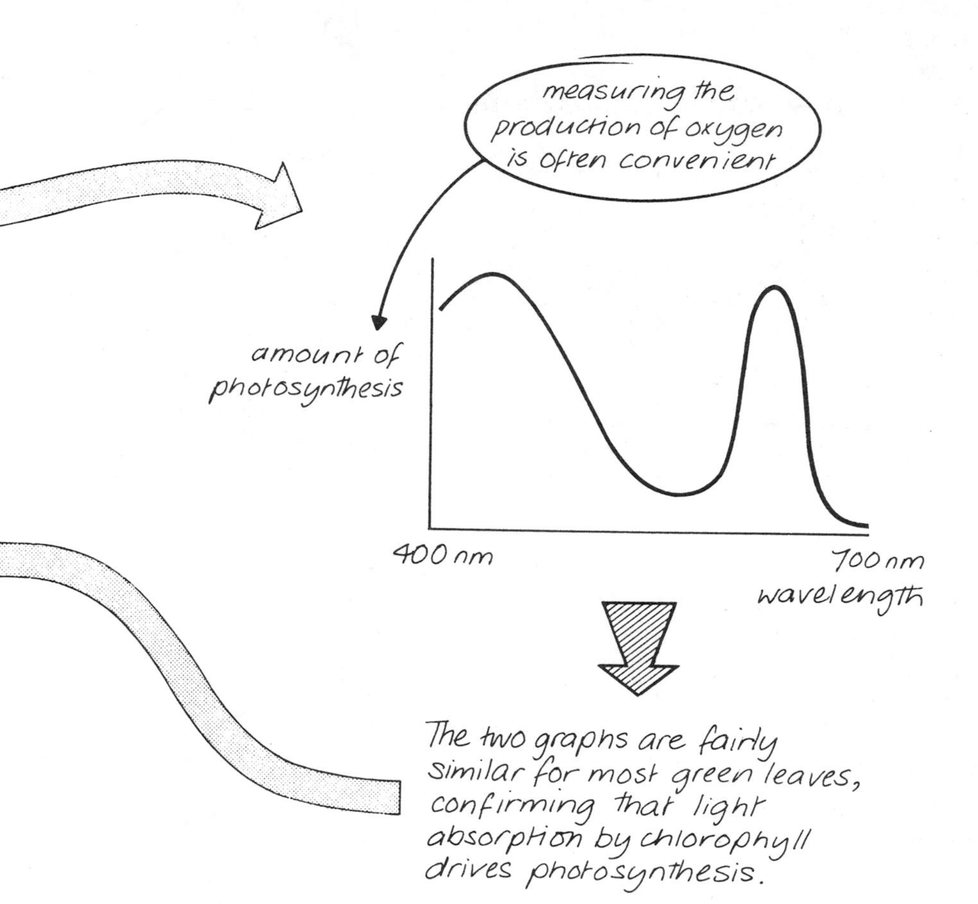

measuring the production of oxygen is often convenient
amount of photosynthesis
400 nm
700 nm
wavelength
The two graphs are fairly similar for most green leaves, confirming that light absorption by chlorophyll drives photosynthesis.

Page 88 summarises the reactions showing that photosynthesis consists basically of the movement of electrons [with associated H^+'s] from water (the **electron donor**) to CO_2.

So

chlorophyll, activated by light, must drive an electron stream.

When a chlorophyll molecule absorbs a blue photon (about 300 kJ per einstein) or a red photon (about 160 kJ per einstein) it ejects an electron from its molecule (it has thousands more). The electron carries the energy with it (like a ball thrown into the air from ground level): it falls back, some of its energy being dissipated as heat, but stays for an instant on a 'ledge' some distance away from its original 'place' in the molecule.

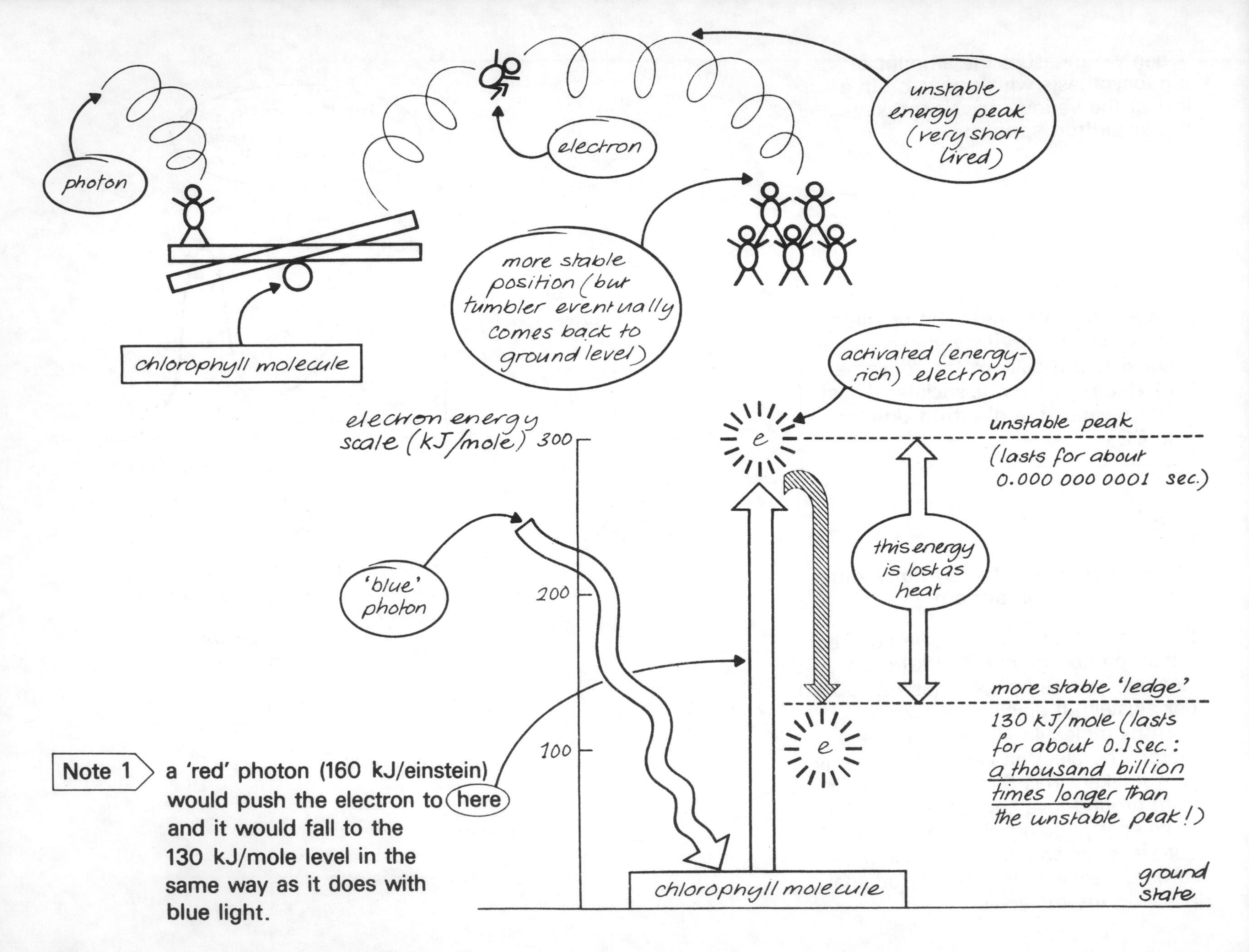
photon
electron
unstable energy peak (very short lived)
chlorophyll molecule
more stable position (but tumbler eventually comes back to ground level)
activated (energy-rich) electron
electron energy scale (kJ/mole)
300
200
100
unstable peak
(lasts for about 0.000 000 0001 sec.)
e
'blue' photon
this energy is lost as heat
more stable 'ledge'
130 kJ/mole (lasts for about 0.1 sec.: a thousand billion times longer than the unstable peak!)
e
chlorophyll molecule
ground state
Note 1
a 'red' photon (160 kJ/einstein) would push the electron to here and it would fall to the 130 kJ/mole level in the same way as it does with blue light.

Note 2 a 'green' photon would not be absorbed [because of the chemical nature of the chlorophyll molecule] even though it would have enough energy to activate the e^- if it were absorbed.

Note 3 an 'infrared' quantum (say 100 kJ/einstein) could not activate e^- as no single quantum would have enough energy to activate any chlorophyll molecule up to 130 kJ/mole – **no matter how much total light falls on the molecule**.

Analogy to Note 3

A child throwing balls (e^-) on to a ledge will not succeed, no matter how much total energy it uses, unless each individual throw (quantum) contains enough energy to lift the ball high enough.

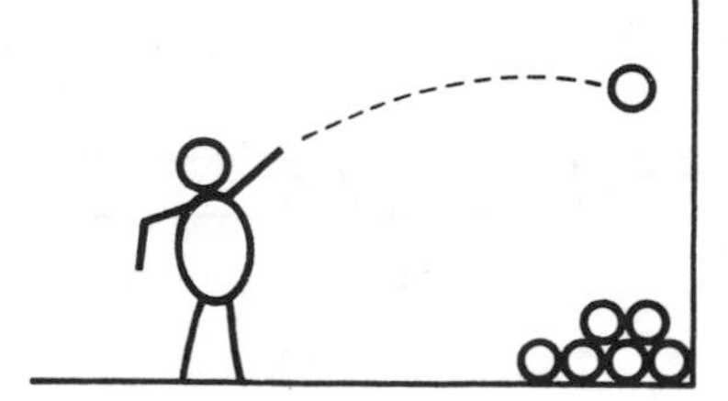

no matter how much energy (how many throws) is expended, the ball will only reach the ledge if an individual throw has enough energy.

Summary so far:

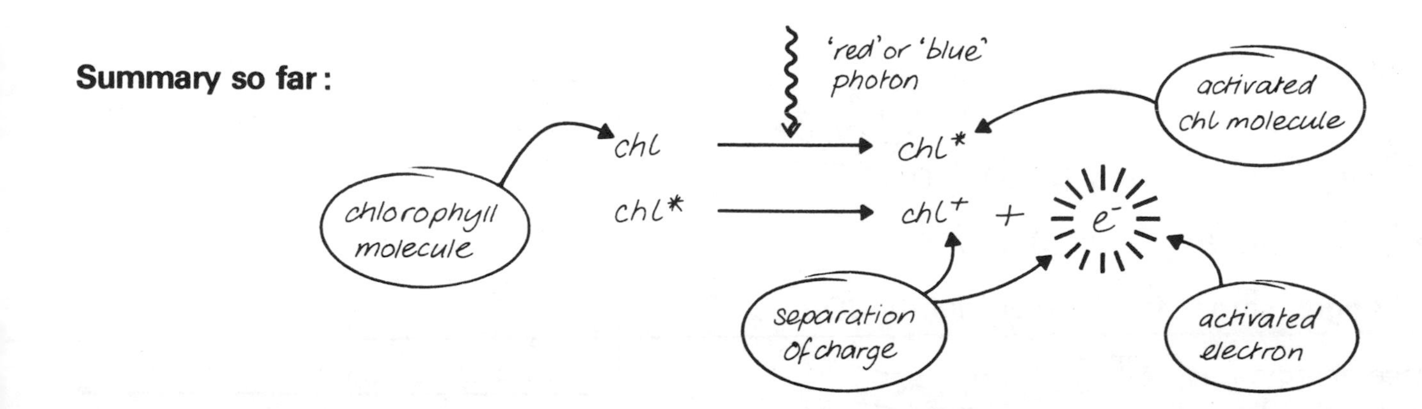

'red' or 'blue' photon
chl → chl*
chl* → chl+ + e-
chlorophyll molecule
activated chl molecule
separation of charge
activated electron

A series of important events now takes place

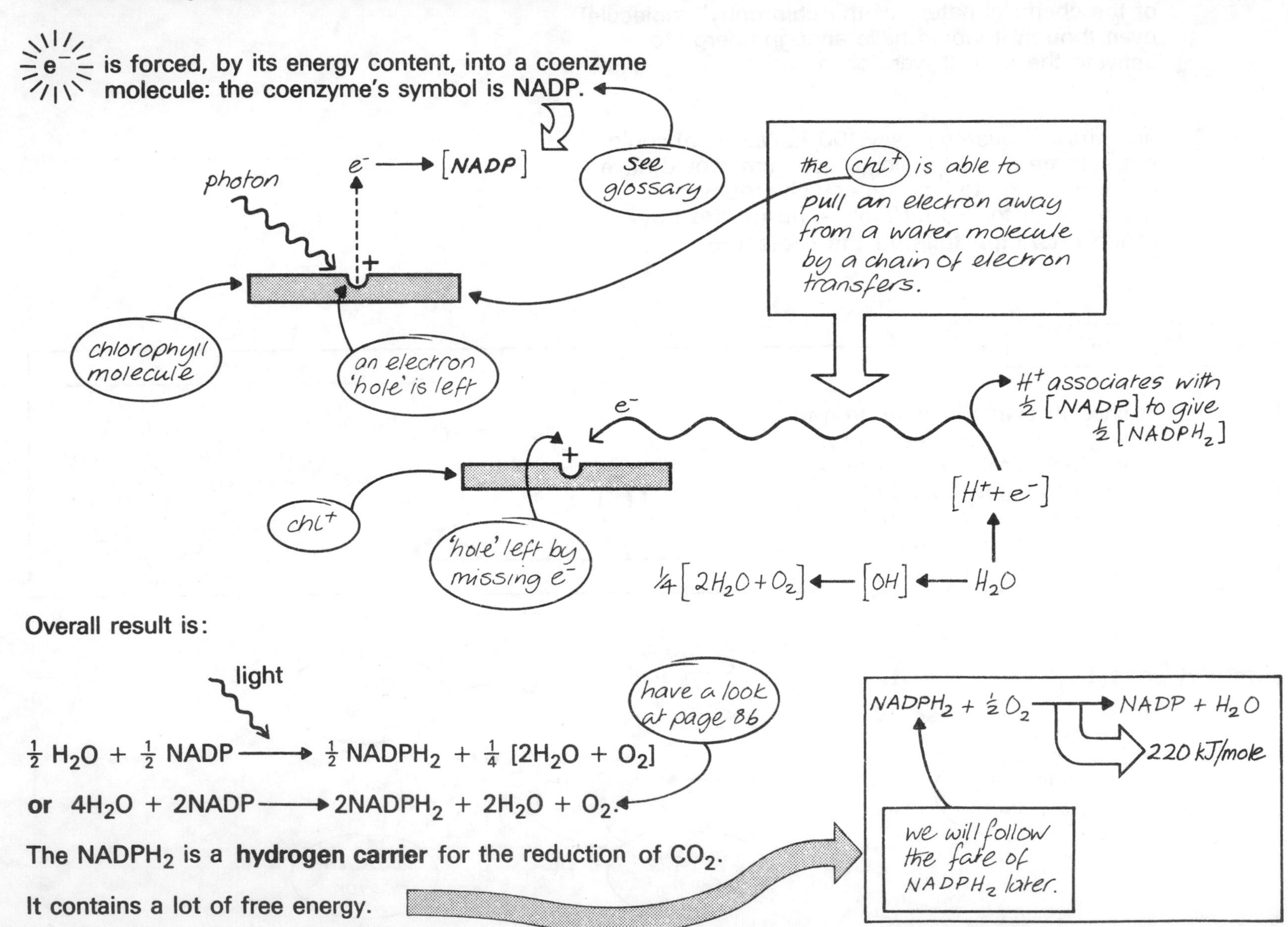

e⁻
is forced, by its energy content, into a coenzyme molecule: the coenzyme's symbol is NADP.
see glossary
photon
e⁻ → [NADP]
chlorophyll molecule
an electron 'hole' is left
the (chl⁺) is able to pull an electron away from a water molecule by a chain of electron transfers.
e⁻
H⁺ associates with ½ [NADP] to give ½ [NADPH₂]
chl⁺
'hole' left by missing e⁻
[H⁺ + e⁻]
¼ [2H₂O + O₂] ← [OH] ← H₂O
light
have a look at page 86
NADPH₂ + ½ O₂ → NADP + H₂O
220 kJ/mole
we will follow the fate of NADPH₂ later.

Overall result is:

$$\frac{1}{2}H_2O + \frac{1}{2}NADP \xrightarrow{\text{light}} \frac{1}{2}NADPH_2 + \frac{1}{4}[2H_2O + O_2]$$

or $4H_2O + 2NADP \longrightarrow 2NADPH_2 + 2H_2O + O_2$.

The $NADPH_2$ is a **hydrogen carrier** for the reduction of CO_2.

It contains a lot of free energy.

A question of energy

The preceding scheme is not quite right. The energy needed to reduce a mole of $NADPH_2$ is 220 kJ (for one molecule we need 2 electrons) which is 110 kJ per mole electron. This is about the same as (actually 85% of) that made available by one mole quantum (i.e. one einstein) absorbed by chlorophyll (see page 96).

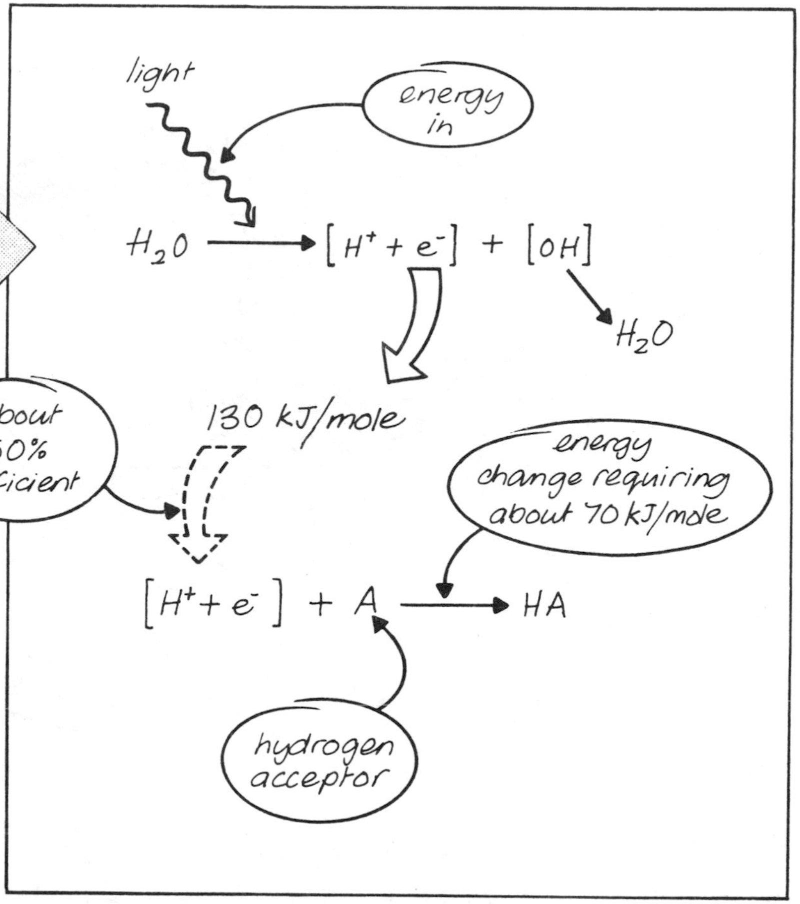

Biological energy transfer reactions (e.g. energy of electrons from chlorophyll ⟶ NADP) are never (? 100%) efficient and commonly have efficiencies of about 40–50%.

The formation of $NADPH_2$ actually requires therefore at least 4 photons.

The absorption of the extra photons is also done by chlorophyll molecules, but they are 'situated' at a different place in the reaction pathway.

Note that we need 4[H] ($2NADPH_2$) for the reduction of one CO_2 molecule.
∴ we need 8 photons for every CO_2 molecule reduced: this is known as **the quantum efficiency** of photosynthesis.

We have

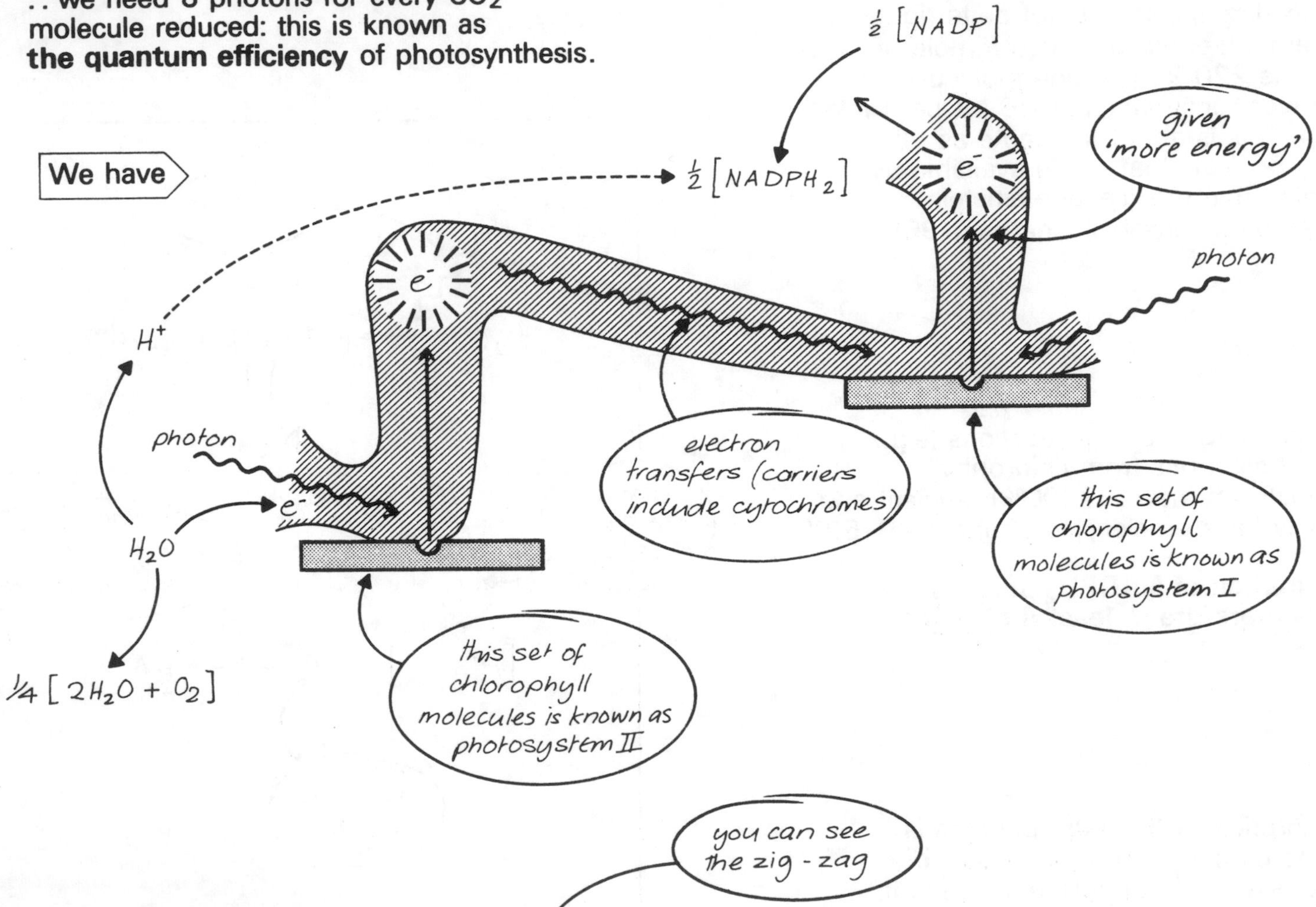

$\frac{1}{2}$ [NADP]
$\frac{1}{2}$ [$NADPH_2$]
given 'more energy'
photon
H^+
photon
e^-
e^-
e^-
H_2O
electron transfers (carriers include cytochromes)
this set of chlorophyll molecules is known as photosystem I
¼ [$2H_2O + O_2$]
this set of chlorophyll molecules is known as photosystem II
you can see the zig - zag

The above is known as 'the Z scheme' in photosynthetic jargon!

Some more notes on the 'Z scheme'

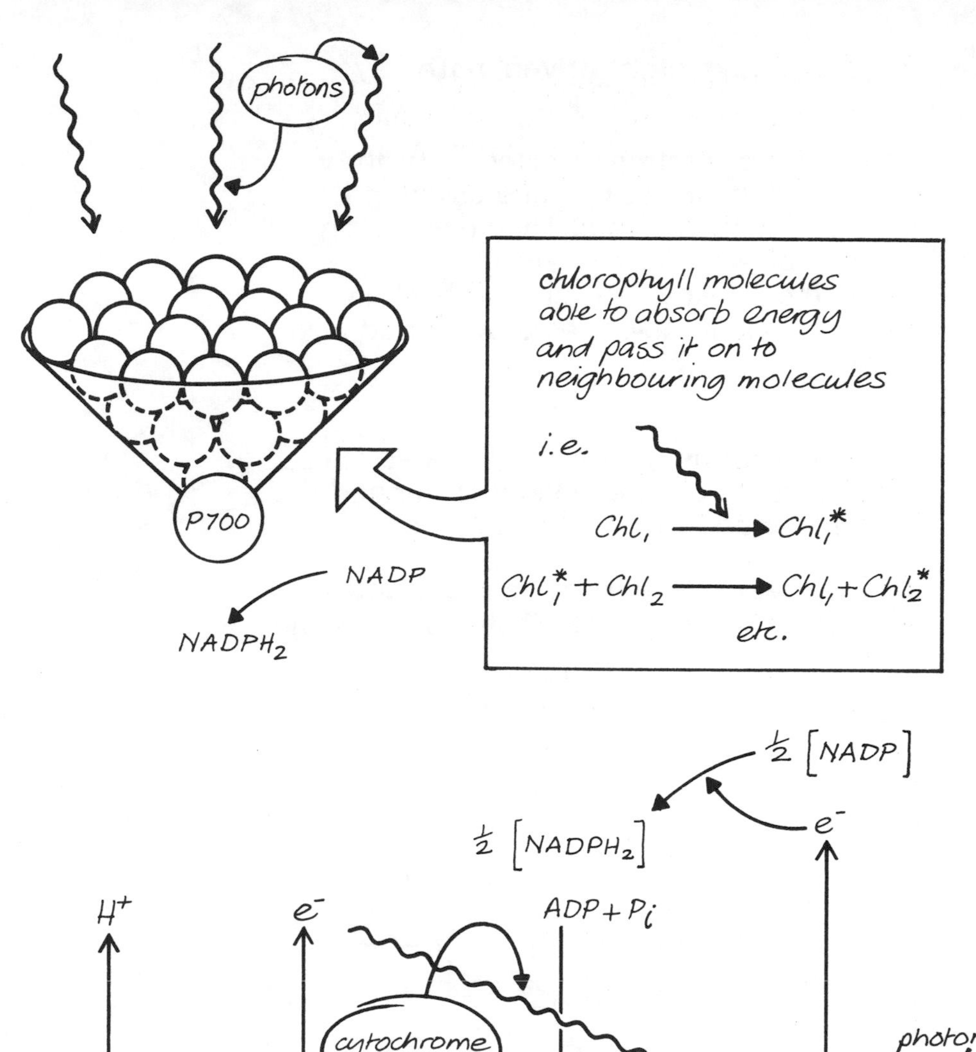
photons
P700
NADP
$NADPH_2$
chlorophyll molecules able to absorb energy and pass it on to neighbouring molecules
i.e.
$Chl_1 \rightarrow Chl_1^*$
$Chl_1^* + Chl_2 \rightarrow Chl_1 + Chl_2^*$
etc.

Note 1

Photosystems I and II actually consist of chlorophyll and possibly other pigments in a regular association or pseudo-crystalline array. This array of pigments is used to 'funnel' energy through one special chlorophyll molecule (called P700) to electron carriers.

Note 2

$NADPH_2$ is not the **only** high energy compound produced. As the electrons pass along the cytochrome chain between photosystem II and photosystem I they lose some of their energy, some of which is trapped as ATP.

This process is called **photophosphorylation**.

When it is associated with 'the Z scheme' it is sometimes referred to as **non-cyclic photophosphorylation**.

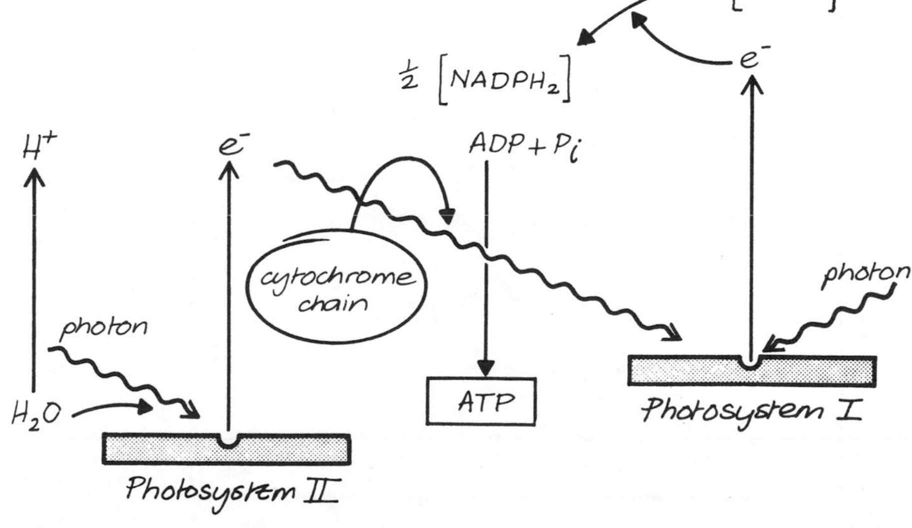
$\frac{1}{2}$ [NADP]
$\frac{1}{2}$ [$NADPH_2$]
e^-
H^+
e^-
ADP + P_i
cytochrome chain
photon
photon
H_2O
ATP
Photosystem II
Photosystem I

An alternative 'light driven' pathway

Some of the electrons ejected from the chlorophyll molecules of photosystem I have an alternative fate. They pass along a series of electron carriers gradually losing their energy and end up back in the chlorophyll molecules they started from.

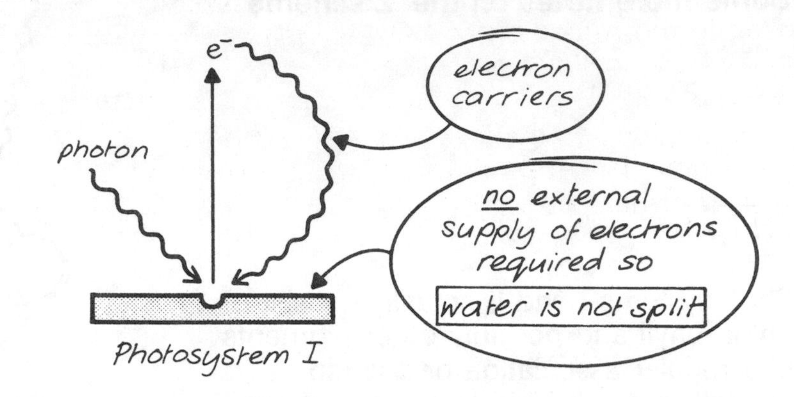

e^-
electron carriers
photon
no external supply of electrons required so
water is not split
Photosystem I

As the electrons pass along the electron carriers some of their energy is trapped in the formation of ATP molecules.

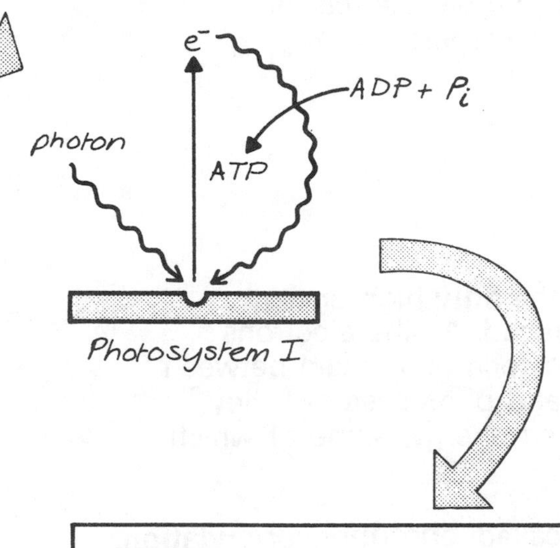

e^-
ADP + P_i
photon
ATP
Photosystem I

This process is known as **cyclic photo-phosphorylation** as the electrons go 'round and round'.

In contrast, the process shown in the Z scheme (page 100) is called non-cyclic photophosphorylation as the electrons do not move cyclically

if

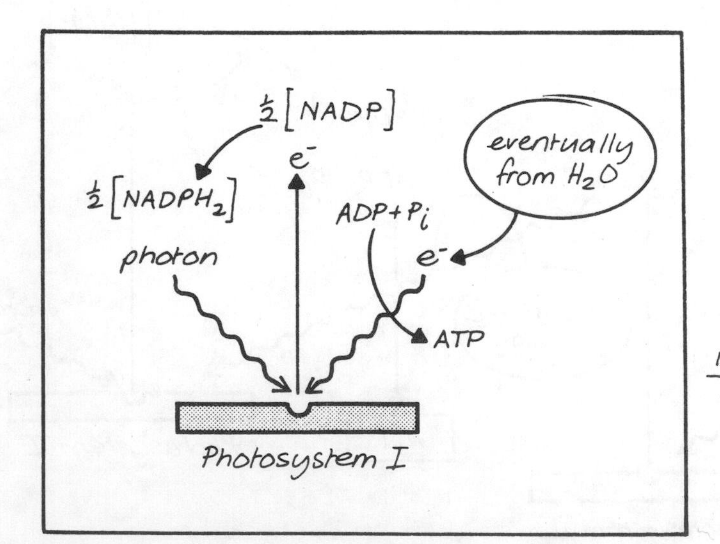

½ [NADP]
½ [$NADPH_2$]
e^-
eventually from H_2O
ADP+P_i
photon
e^-
ATP
Photosystem I

The ATP generated in these two processes is used in the reduction of CO_2 as well as for many other cellular processes.

Production of $NADPH_2$ and ATP in these reactions is a **transduction of light energy to chemical energy.**

Summary

Cyclic flow

Only ATP produced

Water not split

Non-cyclic flow

ATP and $NADPH_2$ produced

Water split to provide [H] for $NADPH_2$

[CHO]

CO_2

$NADPH_2$

Containing energy in the form of reducing power [H]

ATP

part of cyclic flow

part of non-cyclic flow

light

electrons

ATP

I

II

light

contains energy in the form of 'high-energy phosphate' bonds

H_2O

The path of carbon in photosynthesis (The Calvin Cycle)

Remember that CO_2 has to be **reduced** to make carbohydrate. This reduction **does not** take place directly. CO_2 is first picked up by an **acceptor molecule** and the reduction step takes place after this.

As 6 CO_2 molecules are needed to make hexose we can write:

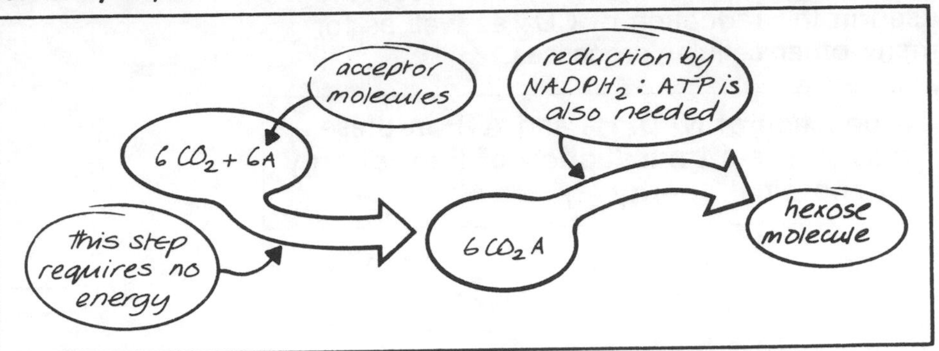

acceptor molecules
reduction by $NADPH_2$: ATP is also needed
$6\ CO_2 + 6A$
this step requires no energy
$6\ CO_2 A$
hexose molecule

What is the acceptor A ?

A is a carbohydrate itself. It is a pentose diphosphate **and is actually one of the products of the chemical reactions involved in CO_2 reduction.**

A is called **ribulose diphosphate**.

Since ribulose diphosphate is, itself, a product of the chemical reactions of photosynthesis we can write:

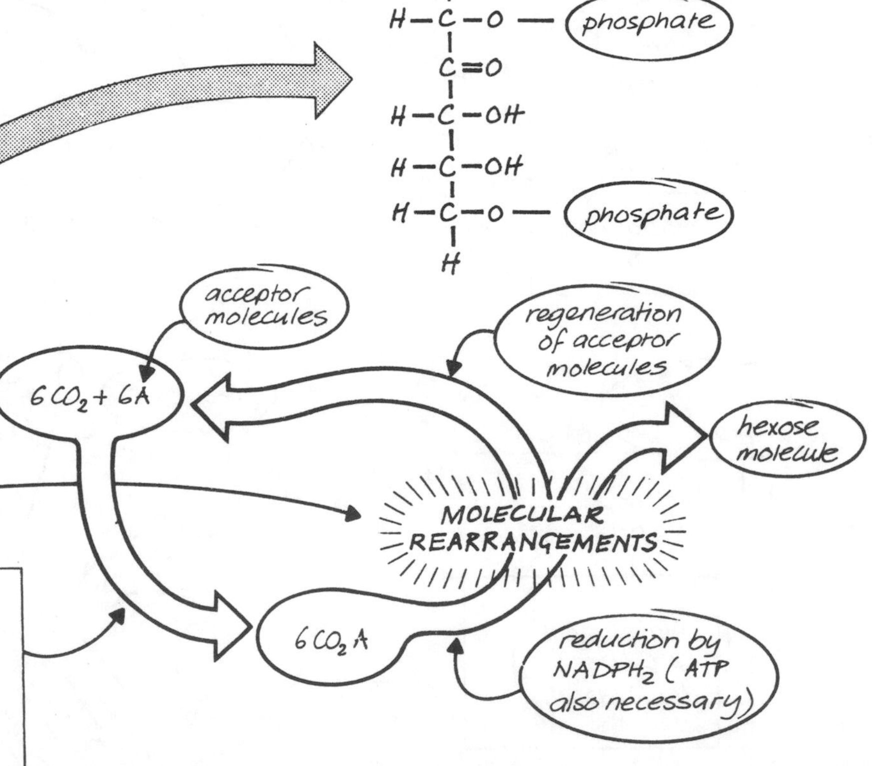

H
H—C—O — phosphate
C=O
H—C—OH
H—C—OH
H—C—O — phosphate
H
acceptor molecules
regeneration of acceptor molecules
$6\ CO_2 + 6A$
hexose molecule
MOLECULAR REARRANGEMENTS
$6\ CO_2 A$
reduction by $NADPH_2$ (ATP also necessary)

During the reduction step and subsequent regeneration of acceptor molecules, the C atoms are 'jumbled up' so that some of the C atoms in the hexose molecule come from CO_2 and some from the ribulose diphosphate.

This cyclic process is known as the Calvin cycle.

This is how it happens:

When a molecule of ribulose diphosphate combines with CO_2 it falls in half to give **two** molecules of a C_3 compound, **phosphoglyceric acid**.

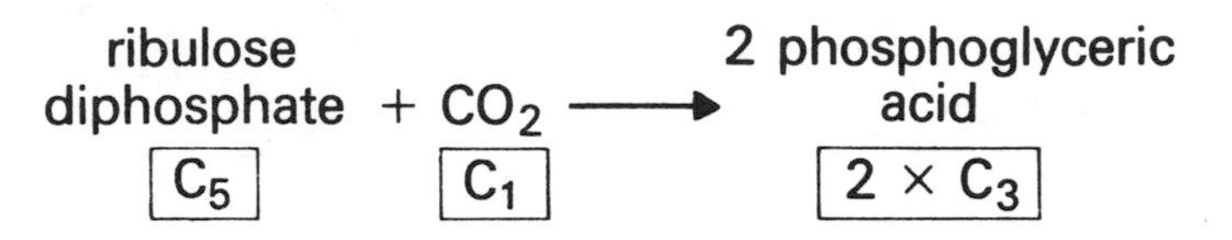

ribulose diphosphate + CO_2 → 2 phosphoglyceric acid
C_5
C_1
$2 \times C_3$

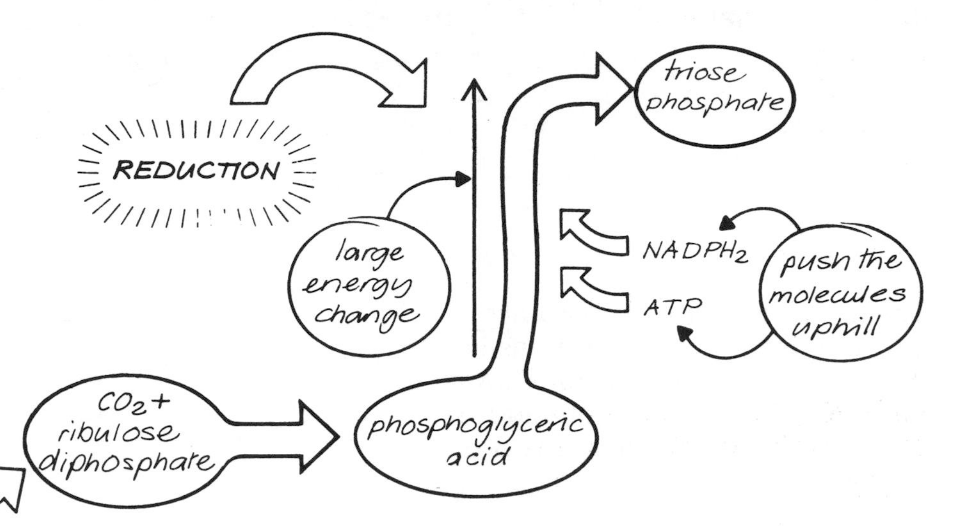

triose phosphate
REDUCTION
large energy change
NADPH2
ATP
push the molecules uphill
CO2 + ribulose diphosphate
phosphoglyceric acid

The reduction step

It is the phosphoglyceric acid which is reduced

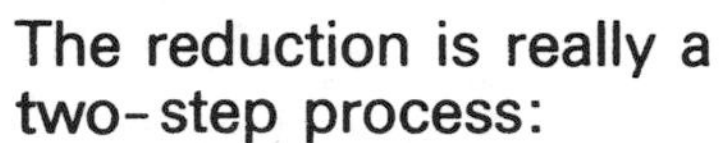

The reduction is really a two-step process:

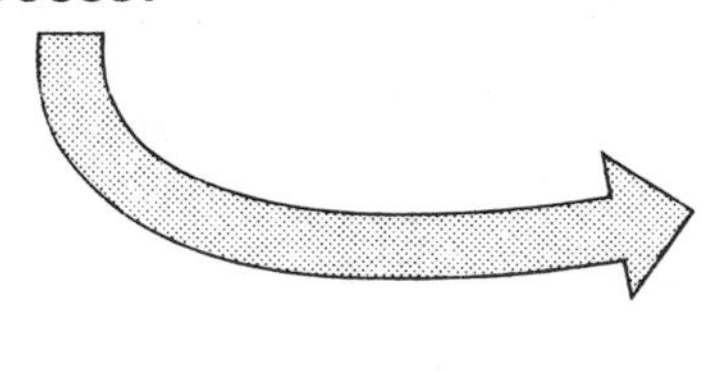

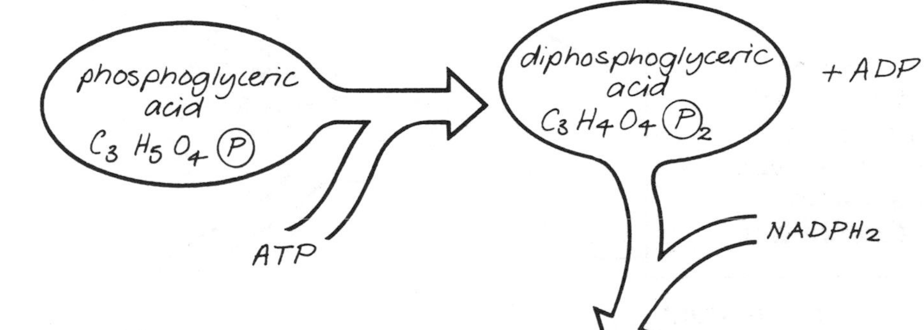

phosphoglyceric acid
$C_3H_5O_4$ (P)
ATP
diphosphoglyceric acid
$C_3H_4O_4$ $(P)_2$
+ ADP
NADPH2

Summary

$C_3H_5O_4$(P) + ATP + $NADPH_2$

⟶ $C_3H_5O_3$(P) + ADP + NADP + P_i + H_2O

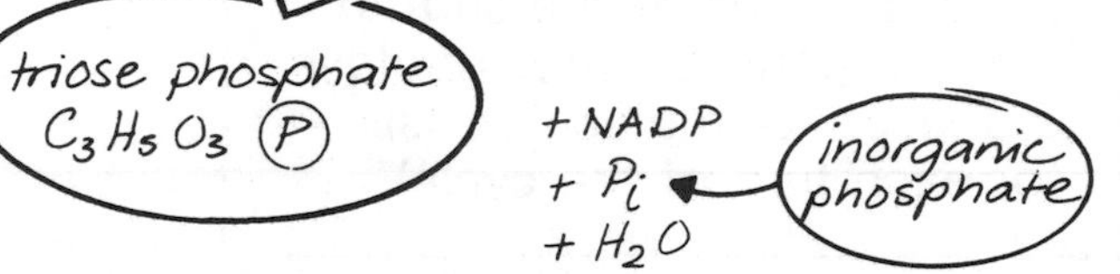

triose phosphate
$C_3H_5O_3$ (P)
+ NADP
+ P_i
+ H_2O
inorganic phosphate

OR considering just the carbon atoms (for an original intake of 6 CO_2 molecules) we have:

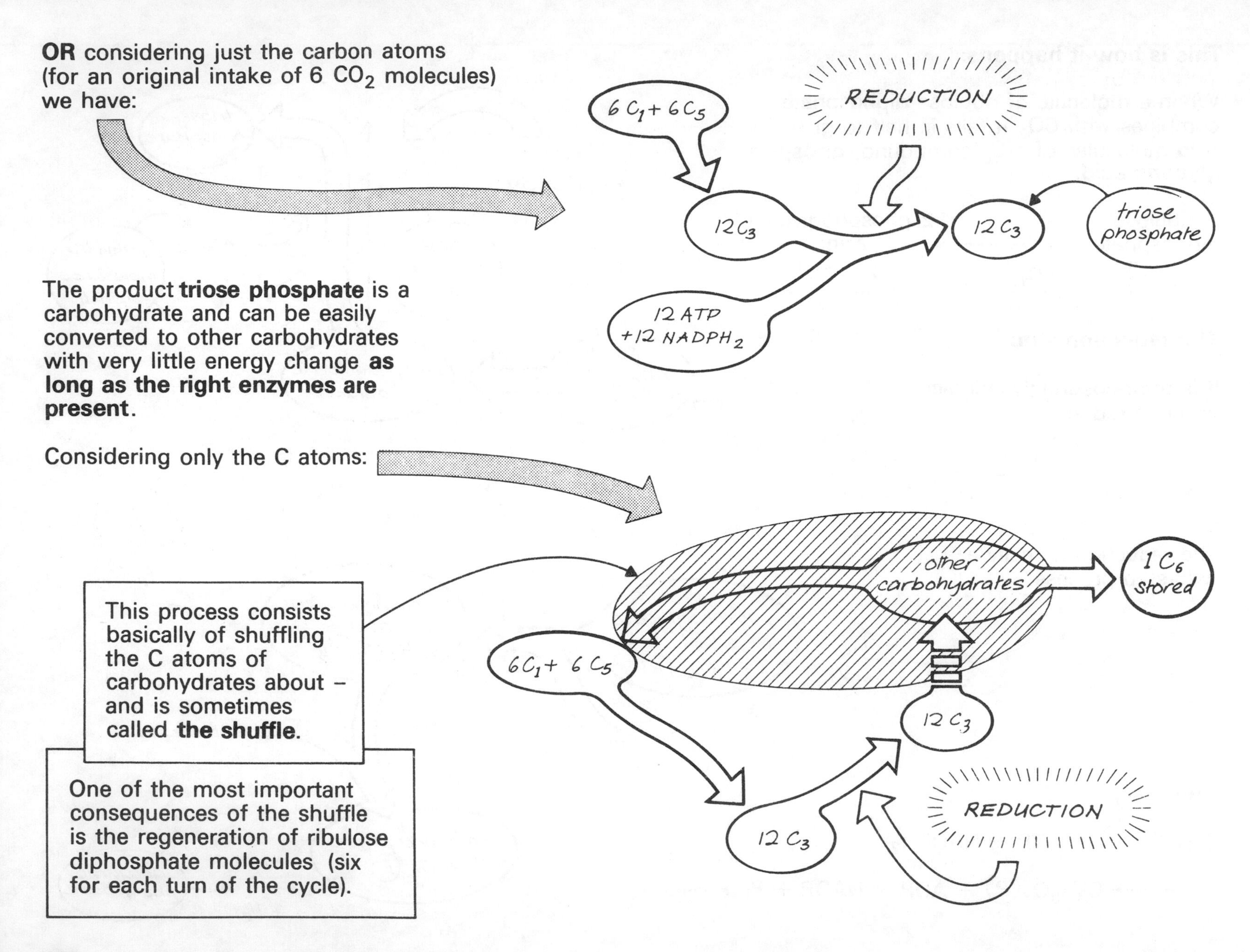

6 C1 + 6 C5
REDUCTION
12 C3
12 C3
triose phosphate
12 ATP + 12 NADPH2
other carbohydrates
1 C6 stored
6 C1 + 6 C5
12 C3
12 C3
REDUCTION

The product **triose phosphate** is a carbohydrate and can be easily converted to other carbohydrates with very little energy change **as long as the right enzymes are present**.

Considering only the C atoms:

This process consists basically of shuffling the C atoms of carbohydrates about – and is sometimes called **the shuffle**.

One of the most important consequences of the shuffle is the regeneration of ribulose diphosphate molecules (six for each turn of the cycle).

The actual path of the C atoms for the production of one hexose unit:

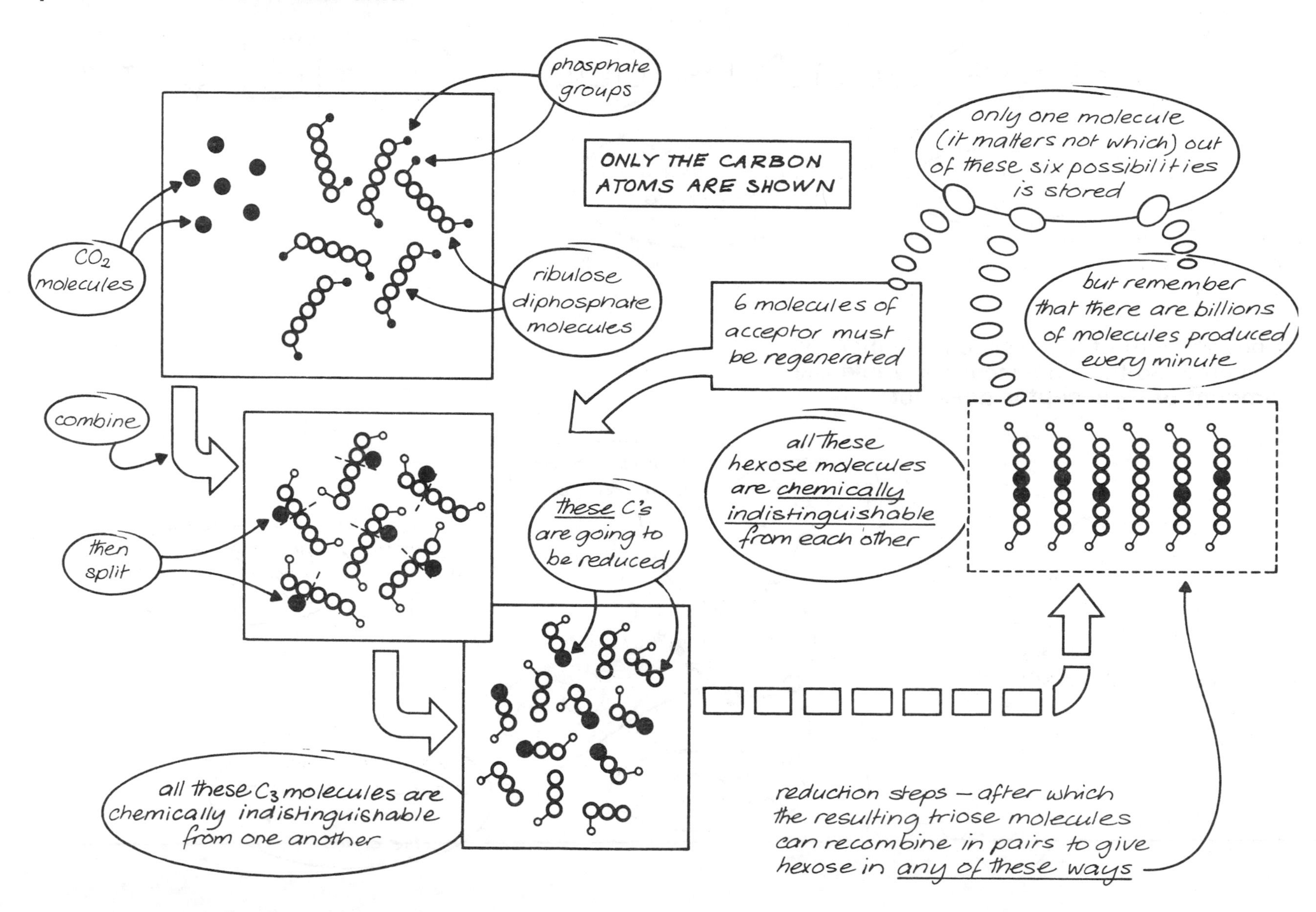

phosphate groups
ONLY THE CARBON ATOMS ARE SHOWN
only one molecule (it matters not which) out of these six possibilities is stored
CO2 molecules
ribulose diphosphate molecules
6 molecules of acceptor must be regenerated
but remember that there are billions of molecules produced every minute
combine
all these hexose molecules are chemically indistinguishable from each other
these C's are going to be reduced
then split
all these C3 molecules are chemically indistinguishable from one another
reduction steps – after which the resulting triose molecules can recombine in pairs to give hexose in any of these ways

Stoichiometric summary

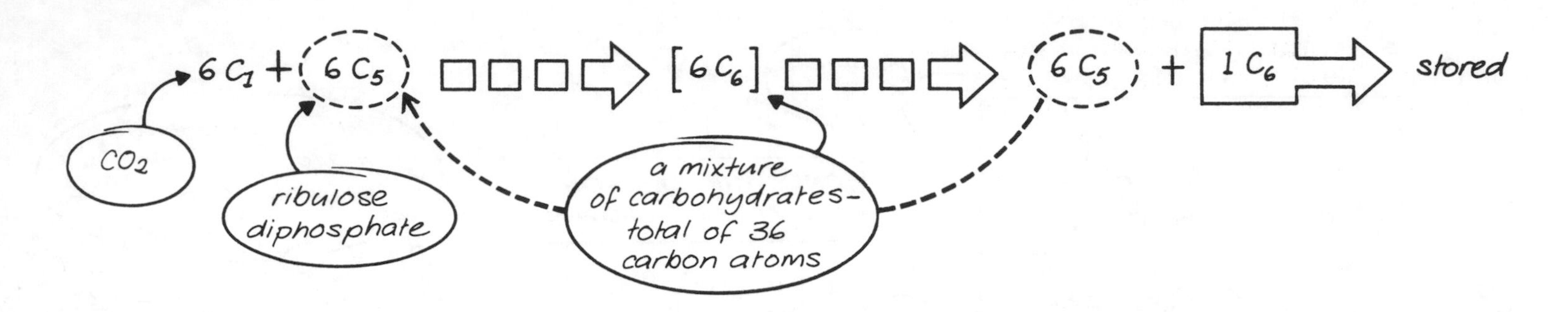

$6C_1$ + $6C_5$ → [$6C_6$] → $6C_5$ + $1C_6$ → stored
CO_2
ribulose diphosphate
a mixture of carbohydrates- total of 36 carbon atoms

For each turn of the cycle one C_6 (hexose) is stored. So the store increases **while the amount of C_5 acceptor does not:**

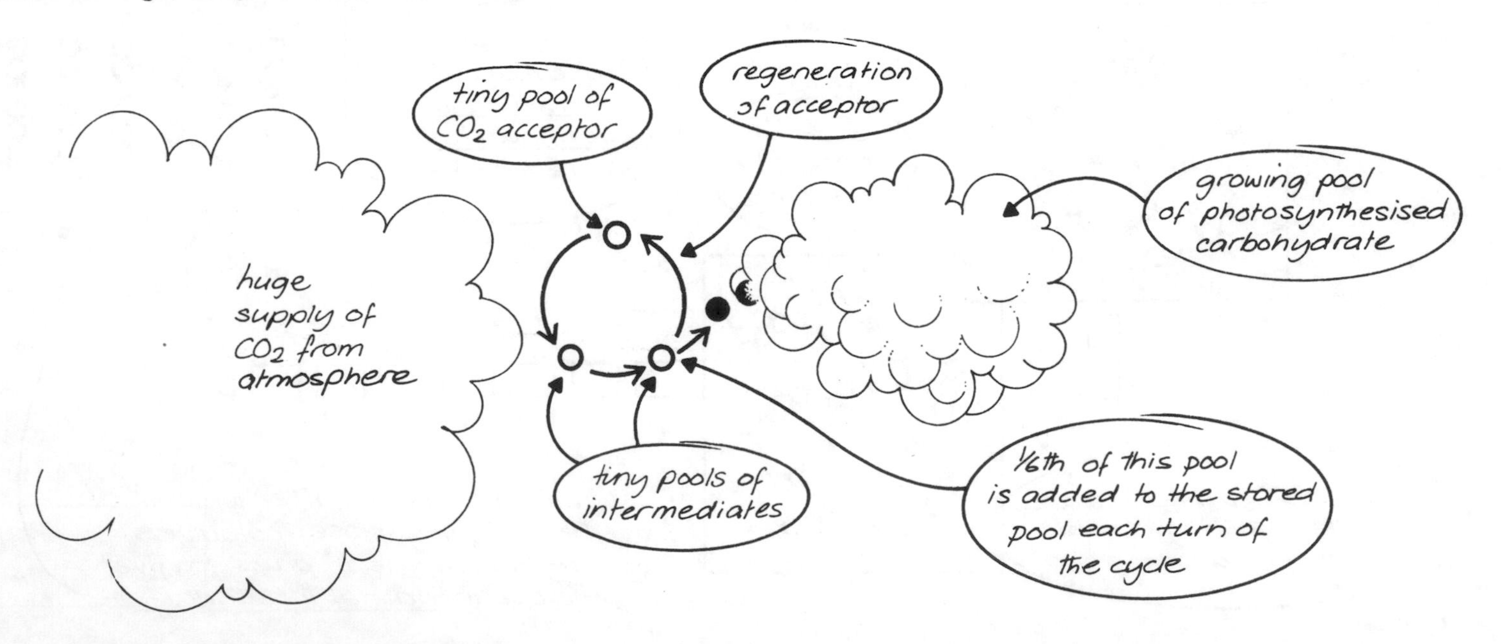

tiny pool of CO_2 acceptor
regeneration of acceptor
growing pool of photosynthesised carbohydrate
huge supply of CO_2 from atmosphere
tiny pools of intermediates
1/6th of this pool is added to the stored pool each turn of the cycle

An alternative method of trapping carbon dioxide

All plants fix carbon dioxide into organic compounds by using the Calvin cycle, but some plants have evolved a method of initially trapping carbon dioxide which is very much more efficient. In these plants the carbon dioxide is trapped in one set of cells from which it is moved into another set of cells to be fixed and converted into carbohydrate. In this way, the cells fixing carbon dioxide are effectively provided with a higher concentration of the gas, which allows them to fix it very much more efficiently (they have a CO_2 'pump').

Trapping the CO_2

The acceptor molecule is **phospho-enol pyruvic acid.**

When a molecule of phospho-enol pyruvic acid combines with CO_2 it produces one molecule of a C_4 compound, **oxaloacetic acid**.

Moving the trapped CO_2

The 'trapped' CO_2 has to be transported to another cell to be finally fixed. To do this the oxaloacetic acid is converted (partly reduced) to **malic acid.**

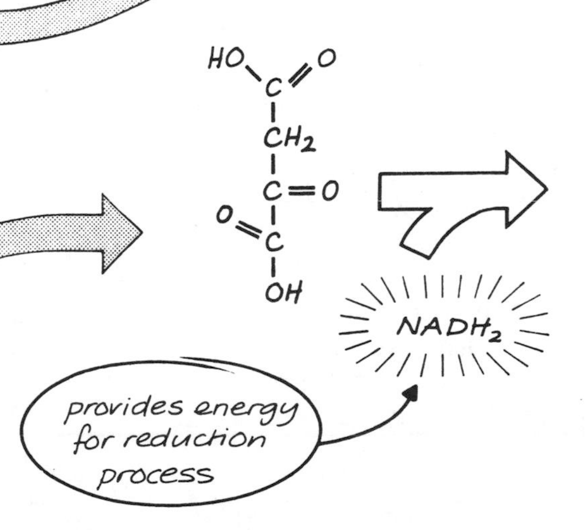
NADH2
provides energy for reduction process

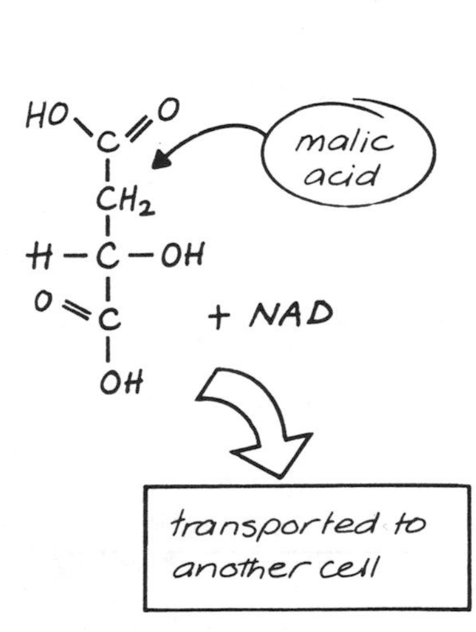
malic acid
+ NAD
transported to another cell

Fixing the CO_2

When the malic acid enters the cell in which final fixation takes place it is first broken down into **CO_2** and **pyruvic acid**.

The released CO_2 is immediately used for carbohydrate production by way of the Calvin cycle. The pyruvate is returned to the first cell for conversion to more acceptor phospho-enol pyruvic acid.

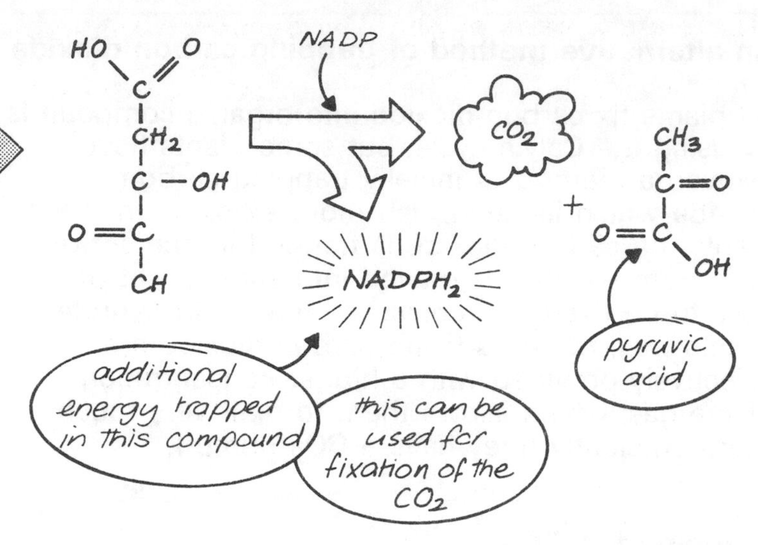

NADP
CO_2
NADPH$_2$
additional energy trapped in this compound
this can be used for fixation of the CO_2
pyruvic acid

We may summarize the process:

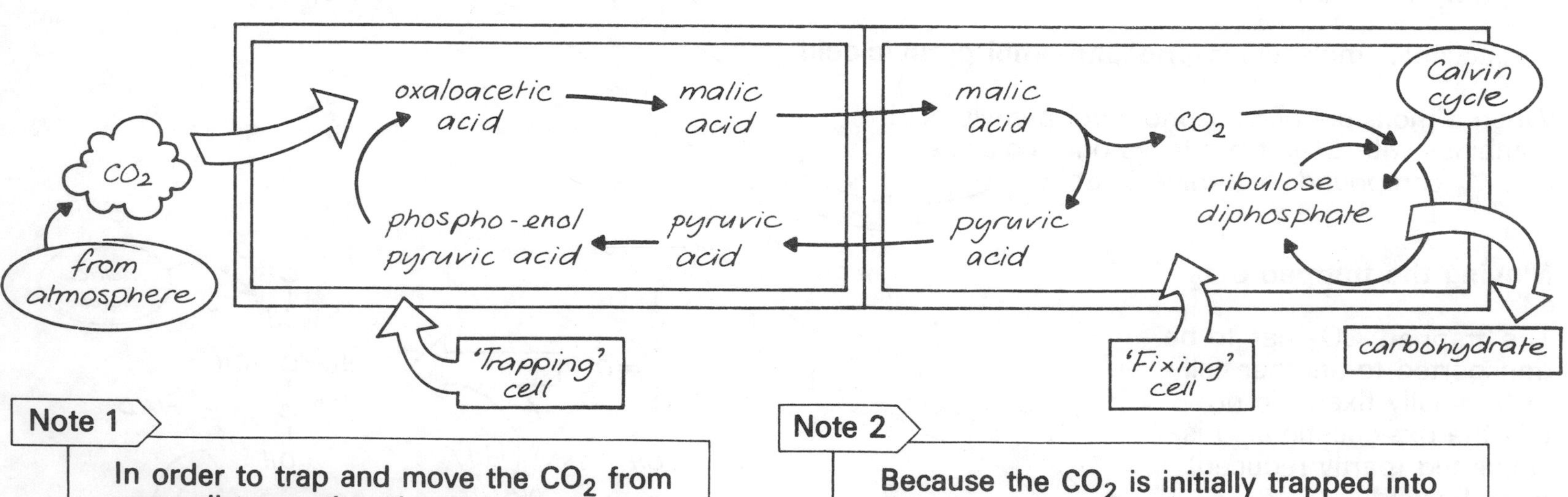

CO_2
from atmosphere
oxaloacetic acid
malic acid
phospho-enol pyruvic acid
pyruvic acid
'Trapping' cell
malic acid
CO_2
Calvin cycle
ribulose diphosphate
pyruvic acid
'Fixing' cell
carbohydrate

Note 1

In order to trap and move the CO_2 from one cell to another the plant has to use extra energy (sunlight) but the increase in efficiency of CO_2 fixation makes this worth while.

Note 2

Because the CO_2 is initially trapped into C_4 compounds the process is sometimes referred to as C_4 photosynthesis and the plants as C_4 plants. Plants using just the Calvin cycle are known as C_3 plants.

There are 2 parts of photosynthesis left to deal with

1. Conversion of hexose phosphate to starch and sucrose (which are the end products of photosynthesis).

2. Regeneration of the molecules (ribulose diphosphate) which accept CO_2.

Sucrose and starch are formed from hexose phosphate without the intermediate formation of free monosaccharide.

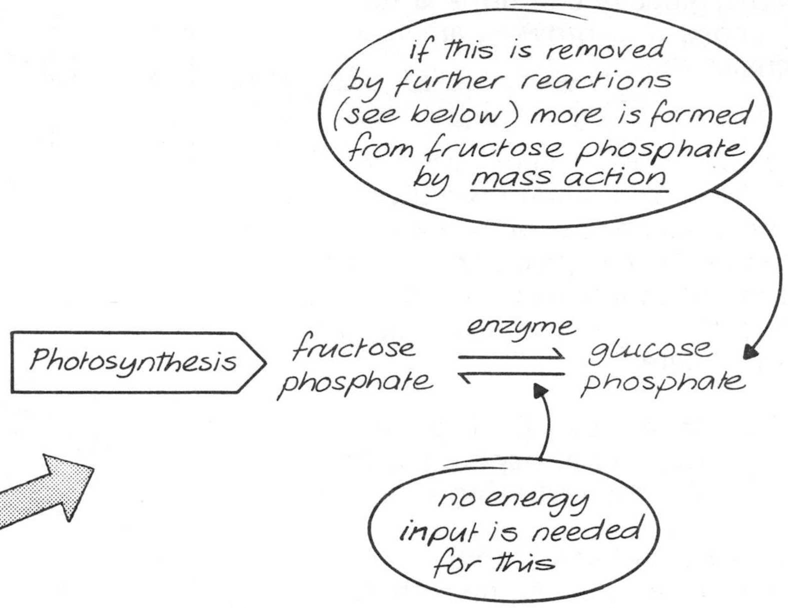

Photosynthesis produces an equilibrium mixture of glucose phosphate and fructose phosphate:

Photosynthesis → fructose phosphate ⇌ (enzyme) glucose phosphate

if this is removed by further reactions (see below) more is formed from fructose phosphate by mass action

no energy input is needed for this

Glucose phosphate can react with ATP (derived from respiration or photosynthesis):

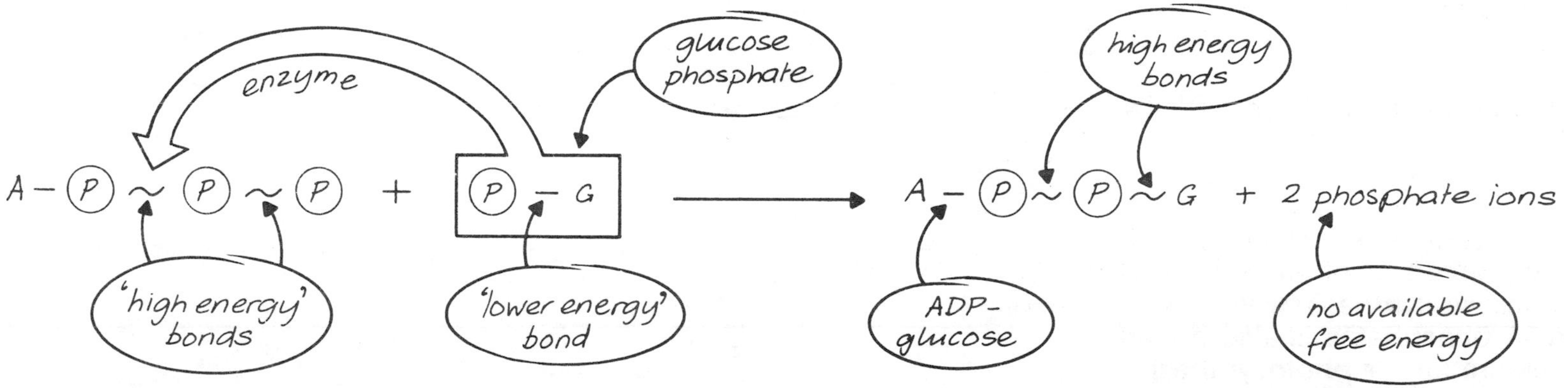

A – (P) ~ (P) ~ (P) + (P) – G ⟶ A – (P) ~ (P) ~ G + 2 phosphate ions

enzyme

glucose phosphate

'high energy' bonds

'lower energy' bond

high energy bonds

ADP-glucose

no available free energy

ADP-glucose can **donate** its glucose to a growing amylose chain:

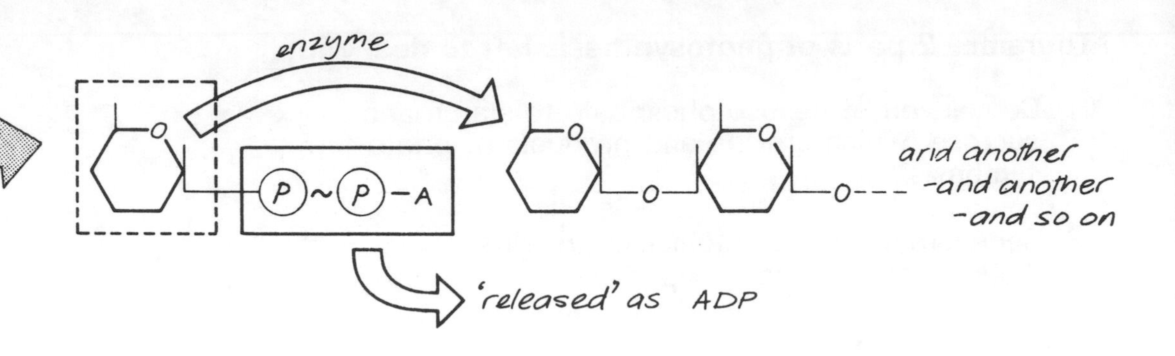
enzyme
O
P ~ P – A
and another
-and another
-and so on
'released' as ADP

The ADP can be converted to ATP by respiration (oxidative phosphorylation), or by photosynthesis (by cyclic or non-cyclic photophosphorylation), and re-used to pick up glucose from glucose phosphate.

So > **starch synthesis from hexose phosphate is an energy requiring process.**

Alternatively to starch synthesis, glucose phosphate may react with a similar compound to ATP called **UTP**.

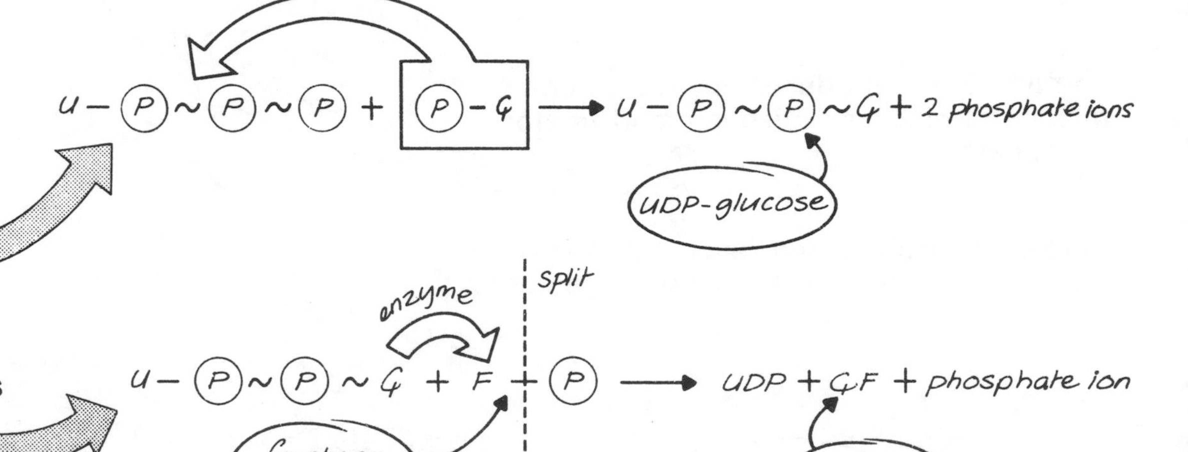
enzyme
U – P ~ P ~ P + P – G → U – P ~ P ~ G + 2 phosphate ions
UDP-glucose
split
enzyme
U – P ~ P ~ G + F + P → UDP + GF + phosphate ion
fructose phosphate
sucrose

Instead of donating its glucose to amylose, UDP–glucose donates it to **fructose-phosphate**.

UTP is regenerated from UDP by ATP.

U – P ~ P + P ~ P ~ P – A → UTP + ADP

enzyme

ADP is itself regenerated to ATP by respiration or photosynthesis. So sucrose synthesis from hexose phosphates is also driven by respiration or photosynthesis.

More about the shuffle

The stoichiometric summary of the scheme shown on page 106 shows that 6 out of a total of 36 carbons (in the form of monosaccharide-phosphates) are stored for each turn of the Calvin cycle. The other 30 **must** be reconverted to 6 C_5 molecules ready to accept another 6 CO_2 molecules. This reconversion, known as the shuffle, takes place with monosaccharide-phosphates — **so no energy is required as all the compounds are on average at the same level of reduction (same energy level).**

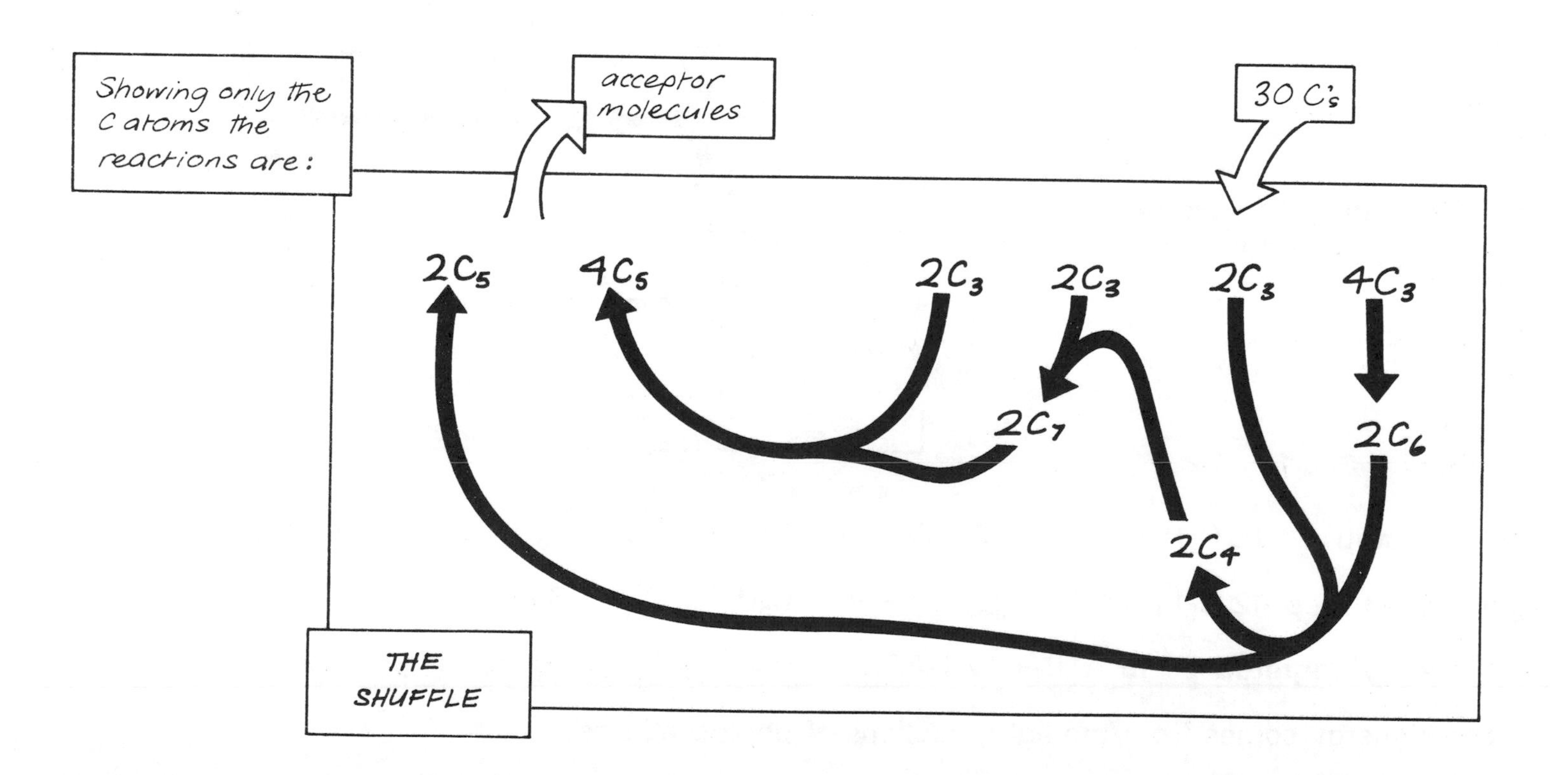
Showing only the C atoms the reactions are:
acceptor molecules
30 C's
2C5
4C5
2C3
2C3
2C3
4C3
2C7
2C6
2C4
THE SHUFFLE

The shuffle reactions produce 6 molecules of a C_5 monophosphate. The final step is to convert these molecules to ribulose diphosphate using ATP.

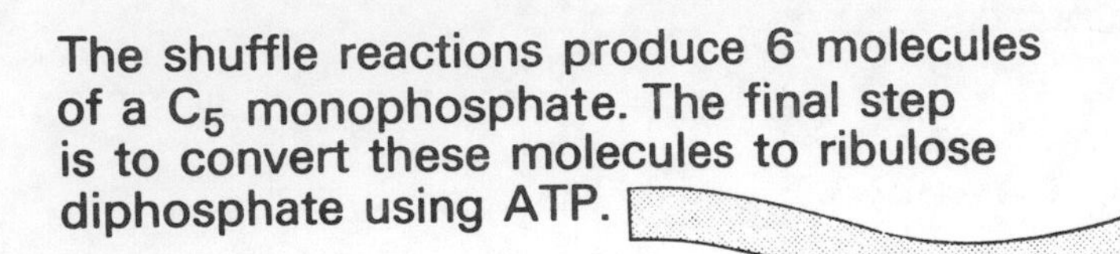

6 pentose monophosphate molecules + 6 ATP ⟶ 6 ribulose diphosphate molecules + 6 ADP

The ATP comes from photosynthetic phosphorylation so light energy drives this part of the Calvin cycle as well.

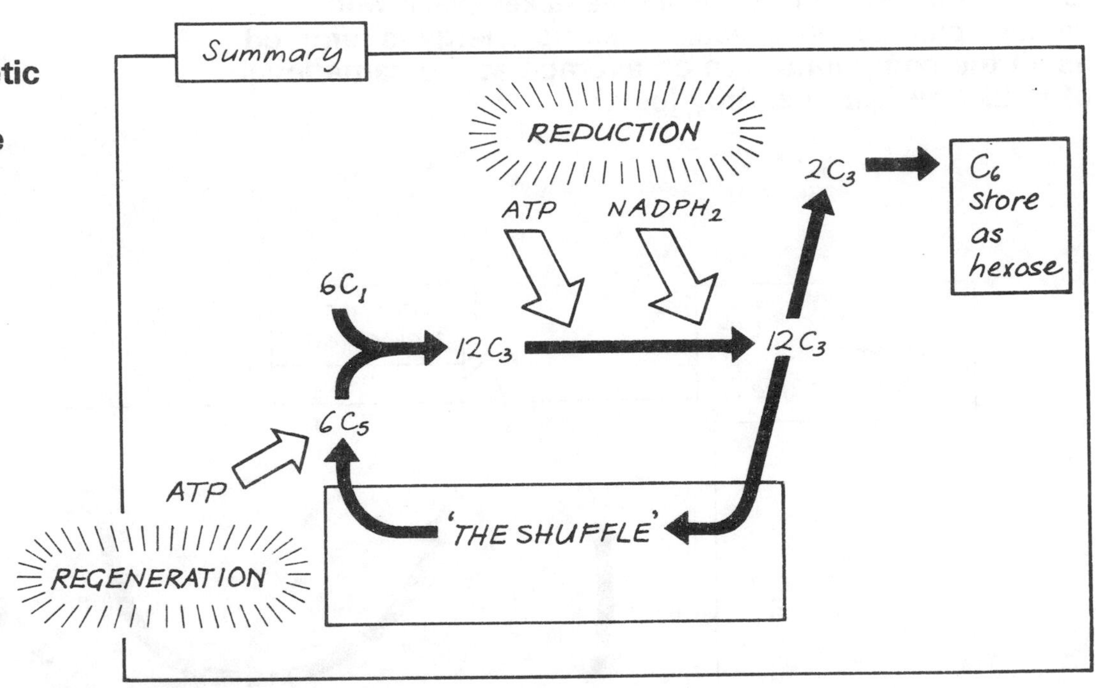

Summary
REDUCTION
ATP
NADPH2
2C3
C6 store as hexose
6C1
12C3
12C3
6C5
ATP
'THE SHUFFLE'
REGENERATION

How much energy do we need for every hexose molecule produced?

Reduction step $12\ C_3H_5O_4(P) + 12\ ATP + 12\ NADPH_2 \longrightarrow 12\ C_3H_5O_3(P) + 12\ NADP + 12\ ADP + 12\ Pi$

Regeneration step $12\ C_5H_9O_5(P) + 6\ ATP \longrightarrow 6\ C_5H_8O_5(P)_2 + 6\ ADP$

Total 'energy' required = 18 ATP + 12 $NADPH_2$ for **every** hexose molecule produced.

All of this energy comes from the light reactions of photosynthesis.

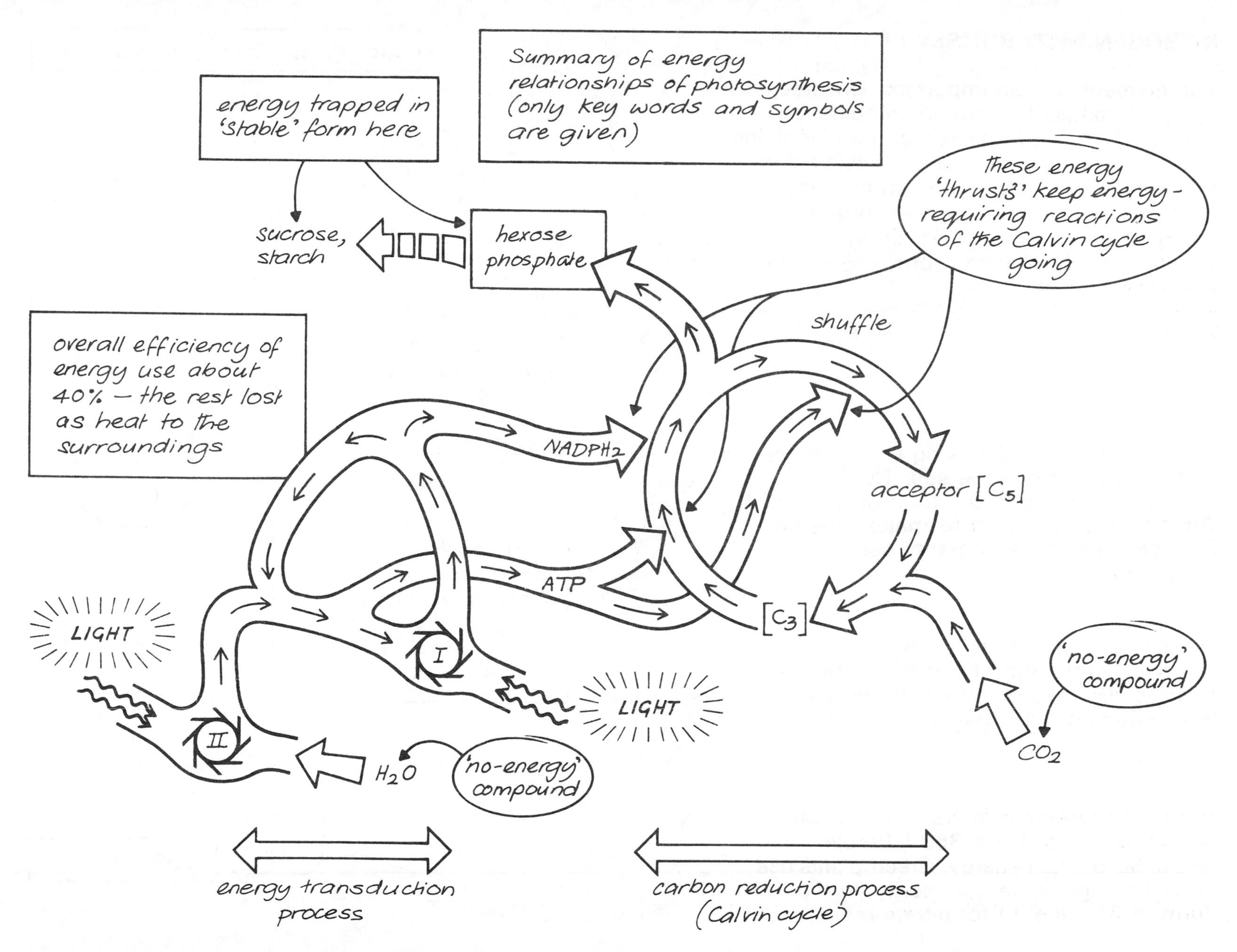

Summary of energy relationships of photosynthesis (only key words and symbols are given)
energy trapped in 'stable' form here
sucrose, starch
hexose phosphate
these energy 'thrusts' keep energy-requiring reactions of the Calvin cycle going
shuffle
overall efficiency of energy use about 40% – the rest lost as heat to the surroundings
NADPH2
acceptor [C5]
ATP
[C3]
LIGHT
I
II
LIGHT
'no-energy' compound
CO2
H2O
'no-energy' compound
energy transduction process
carbon reduction process (Calvin cycle)

NITROGEN METABOLISM

The element N is an important constituent of living things. It occurs in combination with C, H and O, for instance in amino acids (therefore proteins also) and nucleic acids. When in combination with carbon (organic) compounds, N is usually called **organic nitrogen**. Nearly all organic nitrogen is in the **chemically reduced form** and can be considered to be

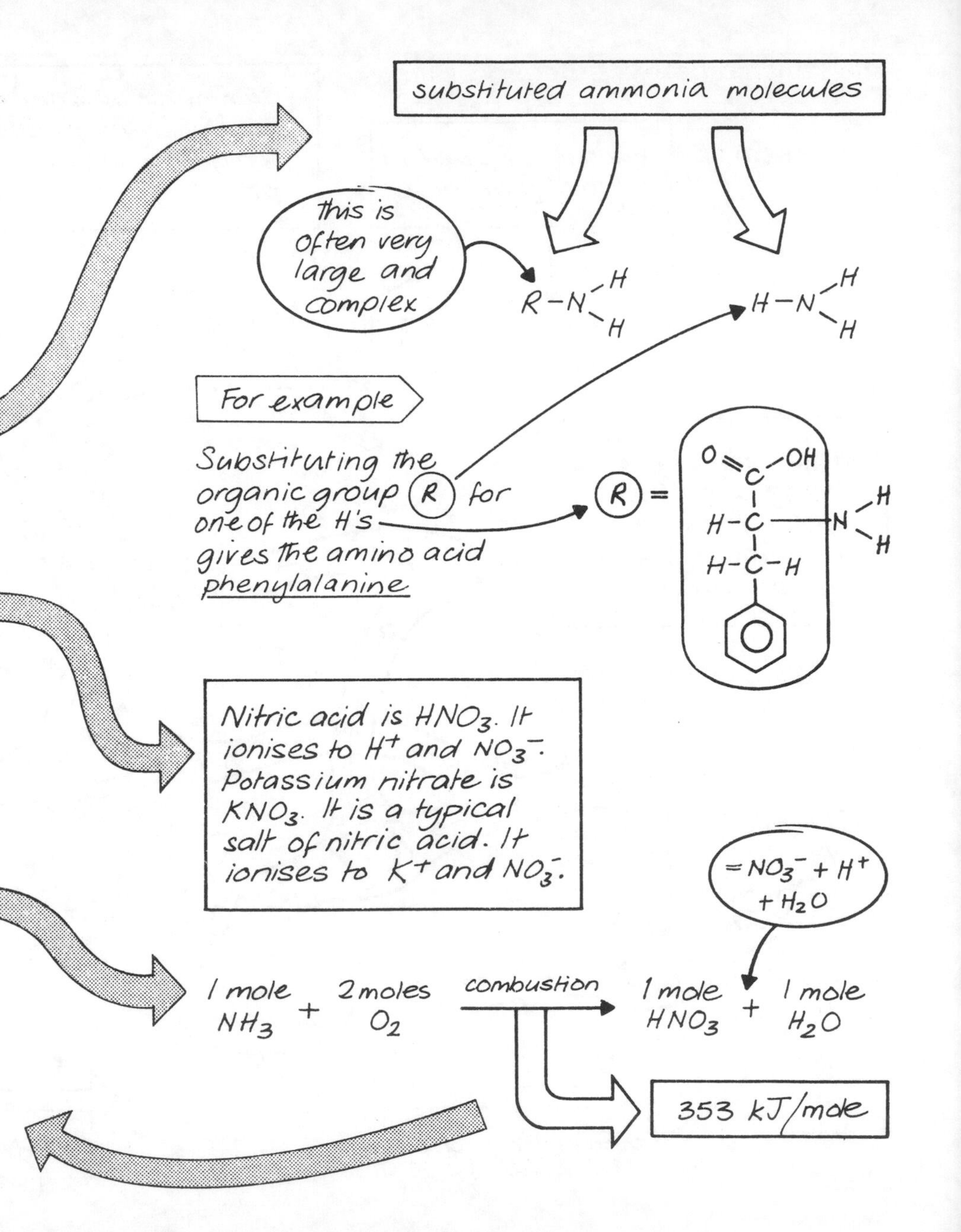

substituted ammonia molecules
this is often very large and complex
R–N–H H
H–N–H H
For example
Substituting the organic group (R) for one of the H's gives the amino acid phenylalanine
(R) =
O=C–OH
H–C–N–H H
H–C–H
Nitric acid is HNO3. It ionises to H+ and NO3−. Potassium nitrate is KNO3. It is a typical salt of nitric acid. It ionises to K+ and NO3−.
= NO3− + H+ + H2O
1 mole NH3 + 2 moles O2 combustion 1 mole HNO3 + 1 mole H2O
353 kJ/mole

Nearly all the N in the living world comes from **nitrate ions in the soil**: NO_3^-. NO_3 is the most oxidised form possible of N. **Animals cannot use it to make organic nitrogen — but green plants can.**

The conversion of NO_3^- to NH_3 is a **chemical reduction. It therefore requires energy: NH_3 has much more energy in the molecule than NO_3^-.**

So any **mechanism** of NO_3^- reduction would require **at least** 353 kJ/mole of available (free) energy. Green plants use sunlight to generate the energy (in the form of $[H^+ + e^-]$) for nitrate reduction.

The **stoichiometry** of the process must be:

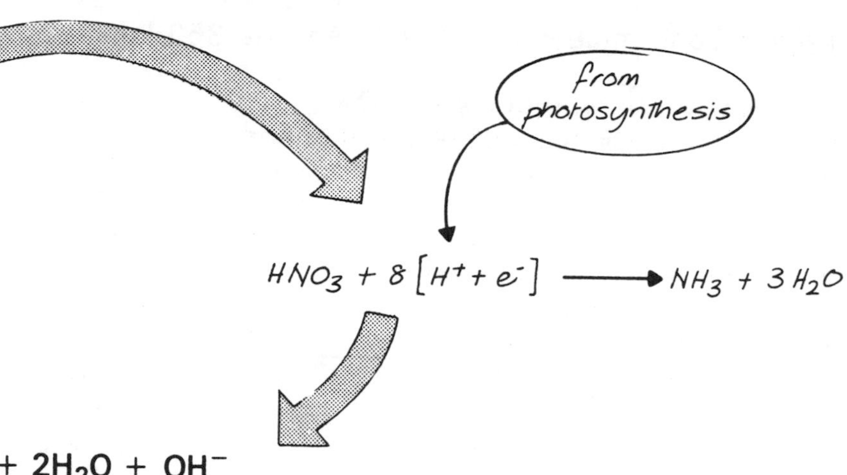

from photosynthesis
$HNO_3 + 8[H^+ + e^-] \longrightarrow NH_3 + 3H_2O$

Reminder

H **ions** $[H^+]$ are not able to reduce NO_3^- (or anything else). It is the **electrons** (of H atoms $[H^+ + e^-]$) which are important.

Which is the same as

$$NO_3^- + 8\,[H] \longrightarrow NH_3 + 2H_2O + OH^-$$
$$(\text{e.g. } KNO_3 + 8\,[H] \longrightarrow NH_3 + 2H_2O + KOH)$$

Some of the [H]'s extract oxygen, the rest add on to N.
Both of these processes are reductions.

The stoichiometry of the process tells us nothing of the pathway.

Much less is known about the pathway of NO_3^- reduction than about CO_2 reduction. We know it takes place in 2 major stages, the first fairly simple, the second much more complex.

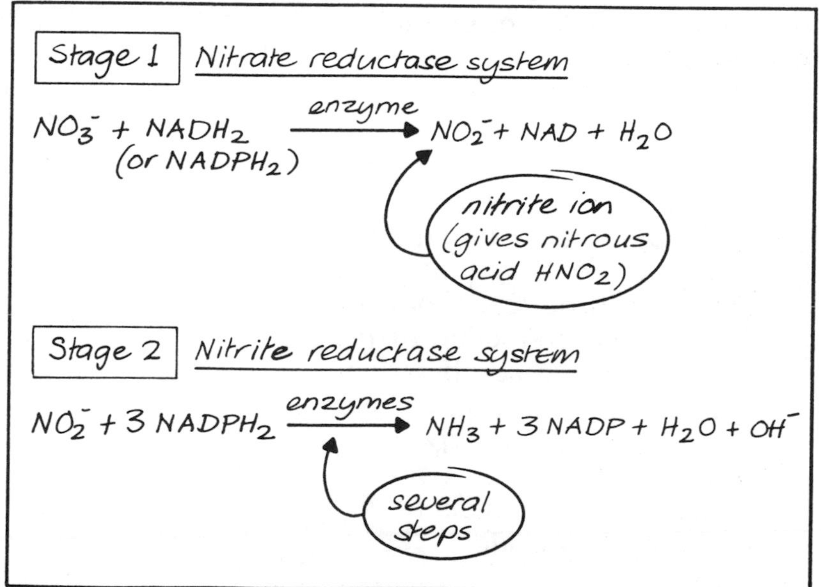

Stage 1
Nitrate reductase system
$NO_3^- + NADH_2$ (or $NADPH_2$) $\xrightarrow{\text{enzyme}}$ $NO_2^- + NAD + H_2O$
nitrite ion (gives nitrous acid HNO_2)
Stage 2
Nitrite reductase system
$NO_2^- + 3\,NADPH_2 \xrightarrow{\text{enzymes}} NH_3 + 3\,NADP + H_2O + OH^-$
several steps

So
4 $NADPH_2$ molecules (or $NADH_2$ molecules which can be derived from $NADPH_2$ molecules by H transfer) are used to reduce one NO_3^- ion.

4 moles of $NADPH_2$ contain:
$4 \times 220 = 880$ kJ of energy.

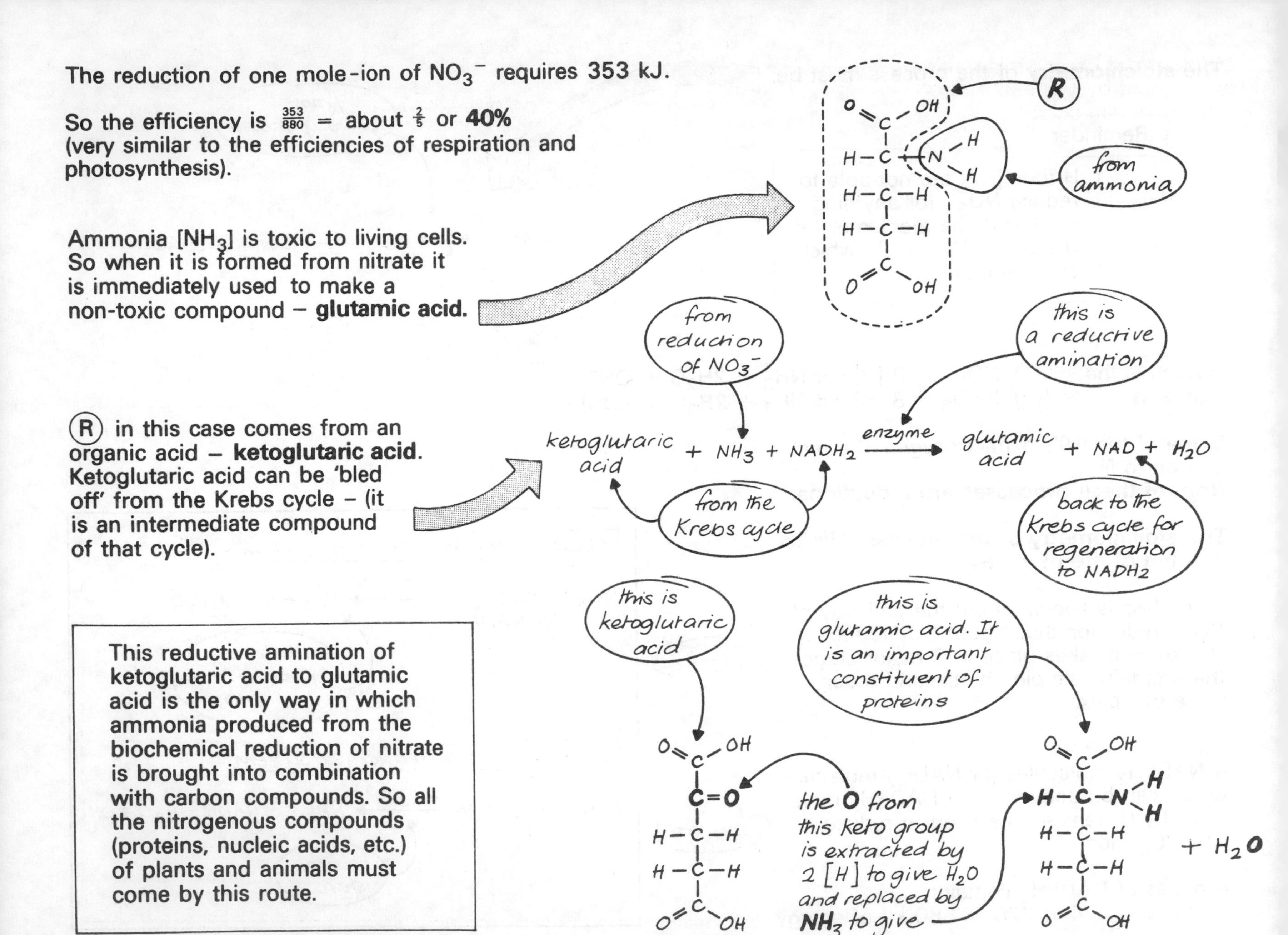

The reduction of one mole-ion of NO_3^- requires 353 kJ.
So the efficiency is $\frac{353}{880}$ = about $\frac{2}{5}$ or **40%** (very similar to the efficiencies of respiration and photosynthesis).
Ammonia [NH_3] is toxic to living cells. So when it is formed from nitrate it is immediately used to make a non-toxic compound – **glutamic acid.**
Ⓡ in this case comes from an organic acid – **ketoglutaric acid**. Ketoglutaric acid can be 'bled off' from the Krebs cycle – (it is an intermediate compound of that cycle).
This reductive amination of ketoglutaric acid to glutamic acid is the only way in which ammonia produced from the biochemical reduction of nitrate is brought into combination with carbon compounds. So all the nitrogenous compounds (proteins, nucleic acids, etc.) of plants and animals must come by this route.
R
from ammonia
from reduction of NO_3^-
this is a reductive amination
ketoglutaric acid + NH_3 + $NADH_2$ → glutamic acid + NAD + H_2O
enzyme
from the Krebs cycle
back to the Krebs cycle for regeneration to $NADH_2$
this is ketoglutaric acid
this is glutamic acid. It is an important constituent of proteins
the **O** from this keto group is extracted by 2 [H] to give H_2O and replaced by NH_3 to give
+ H_2**O**

Many of the other amino acids are formed by **transamination** from glutamic acid:

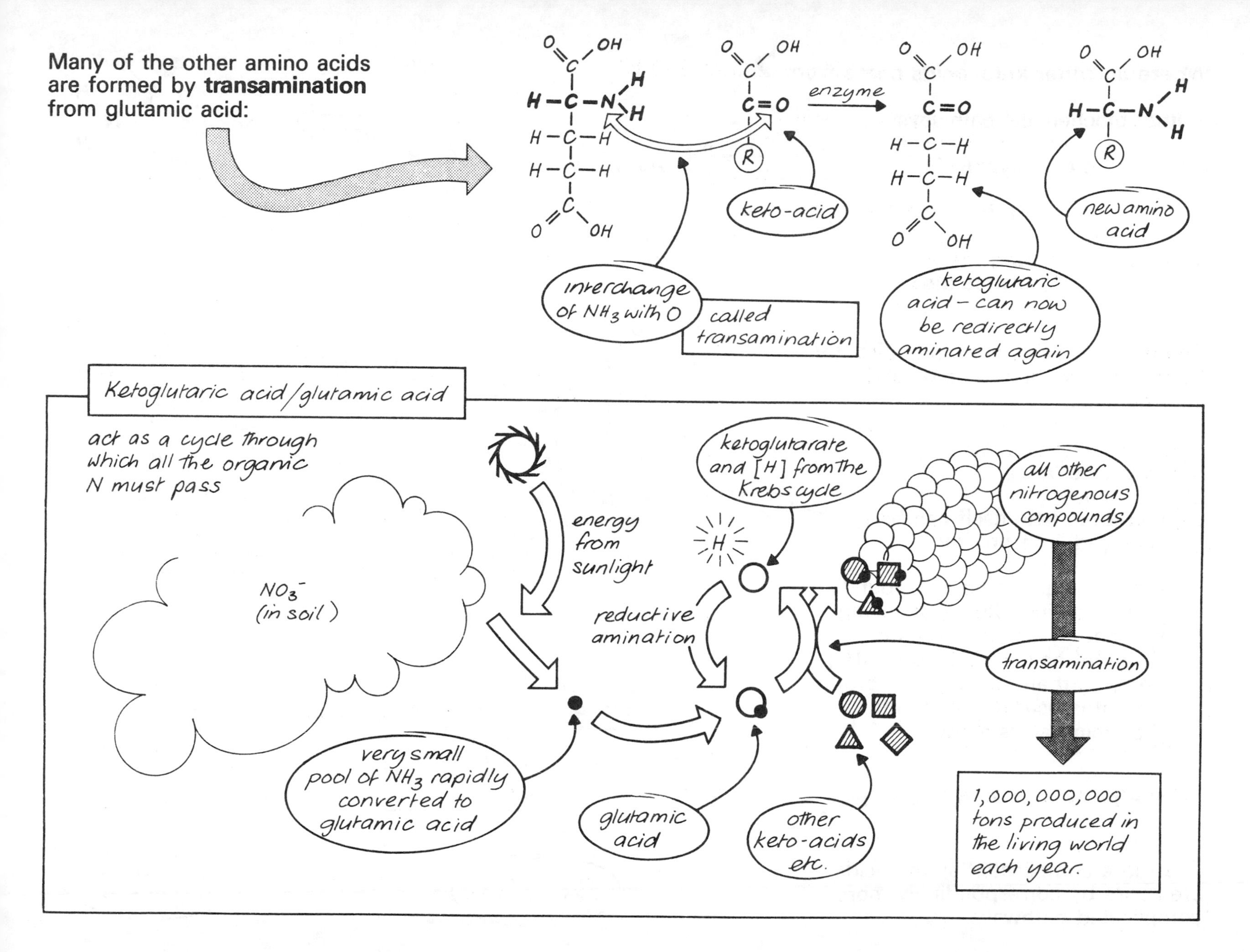
enzyme
keto-acid
new amino acid
interchange of NH_3 with O
called transamination
ketoglutaric acid – can now be redirectly aminated again
Ketoglutaric acid/glutamic acid
act as a cycle through which all the organic N must pass
NO_3^- (in soil)
energy from sunlight
H
ketoglutarate and [H] from the Krebs cycle
all other nitrogenous compounds
reductive amination
transamination
very small pool of NH_3 rapidly converted to glutamic acid
glutamic acid
other keto-acids etc.
1,000,000,000 tons produced in the living world each year.

Where do other keto-acids come from ?

– other biochemical pathways

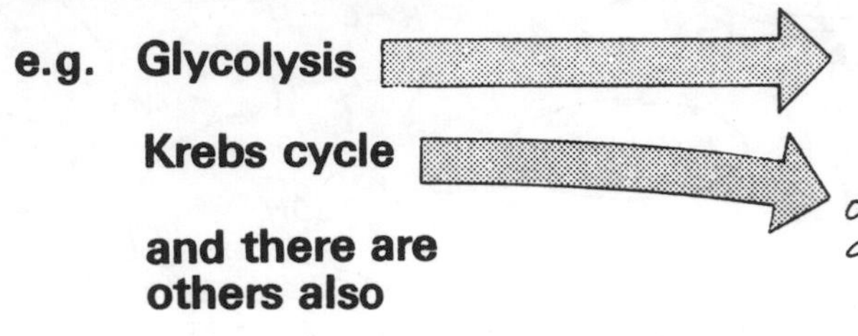

e.g. **Glycolysis**

Krebs cycle

and there are others also

Some amino acids are made **directly** from others without a keto-acid intermediate.

For instance

1. **Glutamic acid** itself can be 'changed' into **proline**.

   Proline is an amino acid found in all proteins. Strictly speaking it is not an amino acid as it does not have an amino ($-NH_2$) group, but an imino (>NH) group –but it is usually put in with the amino acids anyway.

2. **Serine** is converted to **glycine**

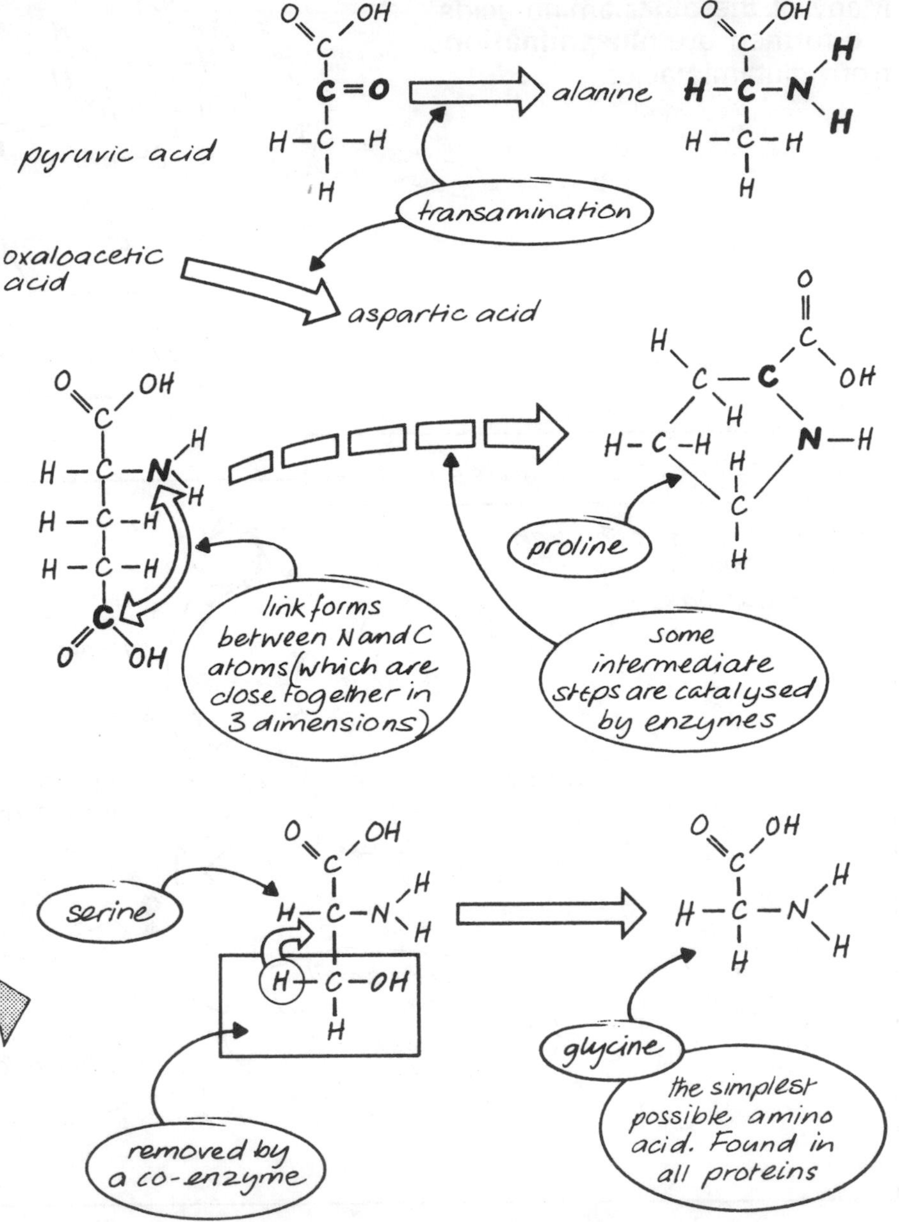

pyruvic acid
alanine
transamination
oxaloacetic acid
aspartic acid
link forms between N and C atoms (which are close together in 3 dimensions)
some intermediate steps are catalysed by enzymes
proline
serine
removed by a co-enzyme
glycine
the simplest possible amino acid. Found in all proteins

The more complicated amino acids are made by correspondingly more complicated pathways.

Nitrogen Fixation

Not all 'organic nitrogen' (nitrogen bound chemically in organic molecules) comes from nitrate in the soil. Some is fixed directly from nitrogen in the air. No higher organisms can do this directly but some plants benefit by associating with nitrogen-fixing bacteria. The best known of these associations is formed between plants of the legume family (clover, beans, etc.) and various bacteria of the genus **Rhizobium**.

For nitrogen to be fixed the triple bond of the N_2 molecule must be broken and three hydrogen atoms bound to each N atom to produce ammonia. The ammonia is then incorporated directly into amino acids. The fixation takes place in root nodules which are formed as a result of the association between the bacteria and plant.

The plant provides carbohydrate and organic acids which are used by the bacteria for their own growth. The bacteria contain a special enzyme called **nitrogenase** which catalyses the reduction of nitrogen to ammonia. The ammonia is used to make amino acids which are utilised by the plant and the bacteria.

We may summarize the process:

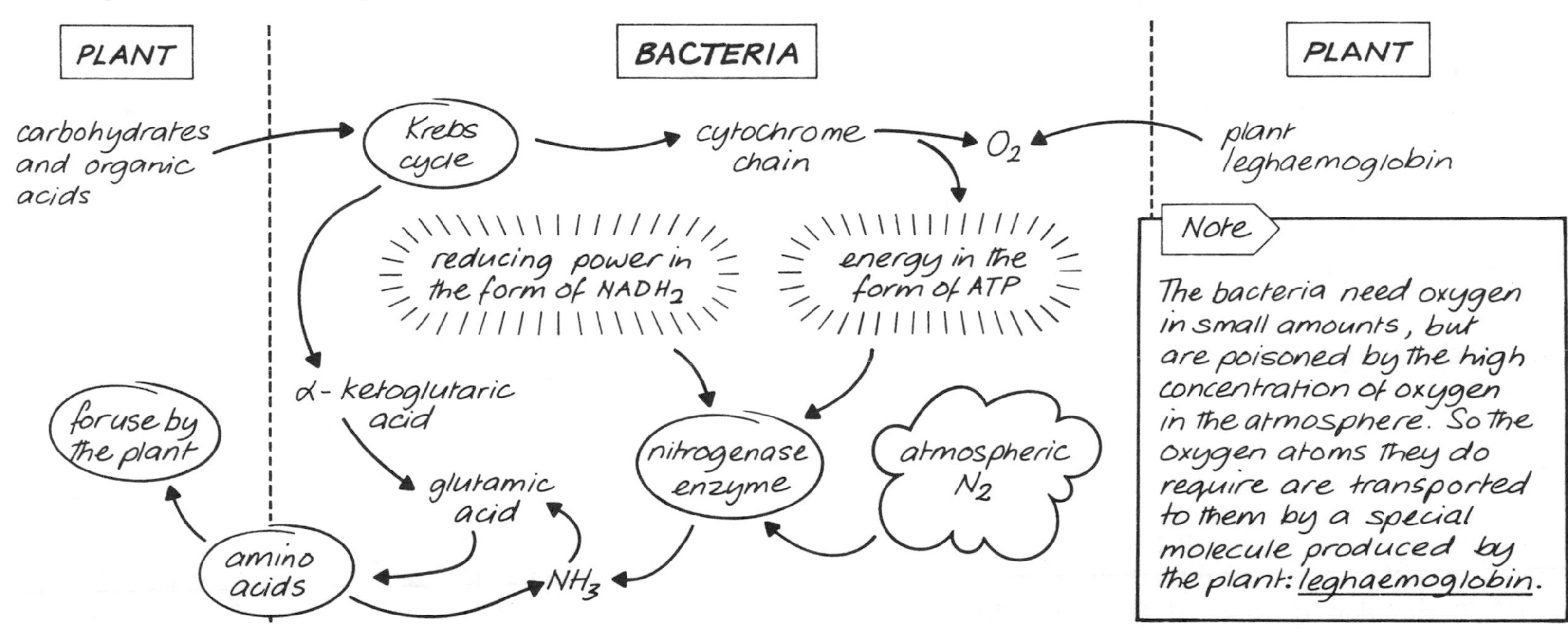
PLANT
BACTERIA
PLANT
carbohydrates and organic acids
Krebs cycle
cytochrome chain
O_2
plant leghaemoglobin
reducing power in the form of $NADH_2$
energy in the form of ATP
α-ketoglutaric acid
glutamic acid
nitrogenase enzyme
atmospheric N_2
for use by the plant
amino acids
NH_3
Note
The bacteria need oxygen in small amounts, but are poisoned by the high concentration of oxygen in the atmosphere. So the oxygen atoms they do require are transported to them by a special molecule produced by the plant: leghaemoglobin.

Excretion of nitrogen

When the proteins and other N compounds are broken down in animal cells, the C is excreted as CO_2 (by the lungs), the H as H_2O (lungs, urine, sweat).

What happens to the N ?

It is excreted **either** as NH_3, **or** as urea, **or** as uric acid.

But > NH_3 is toxic. It damages tissue if its concentration rises above a very low level. Some marine animals excrete NH_3 because it can be diluted in a very large volume of water (the sea!). Land animals convert their excreted N into a safer compound – **either** urea **or** uric acid – which can rise to rather higher concentrations in their bodies without harm.

Man excretes urea. Urea is a simple organic compound of nitrogen.

It is made from ammonia and CO_2. The **stoichiometry** of its formation is:

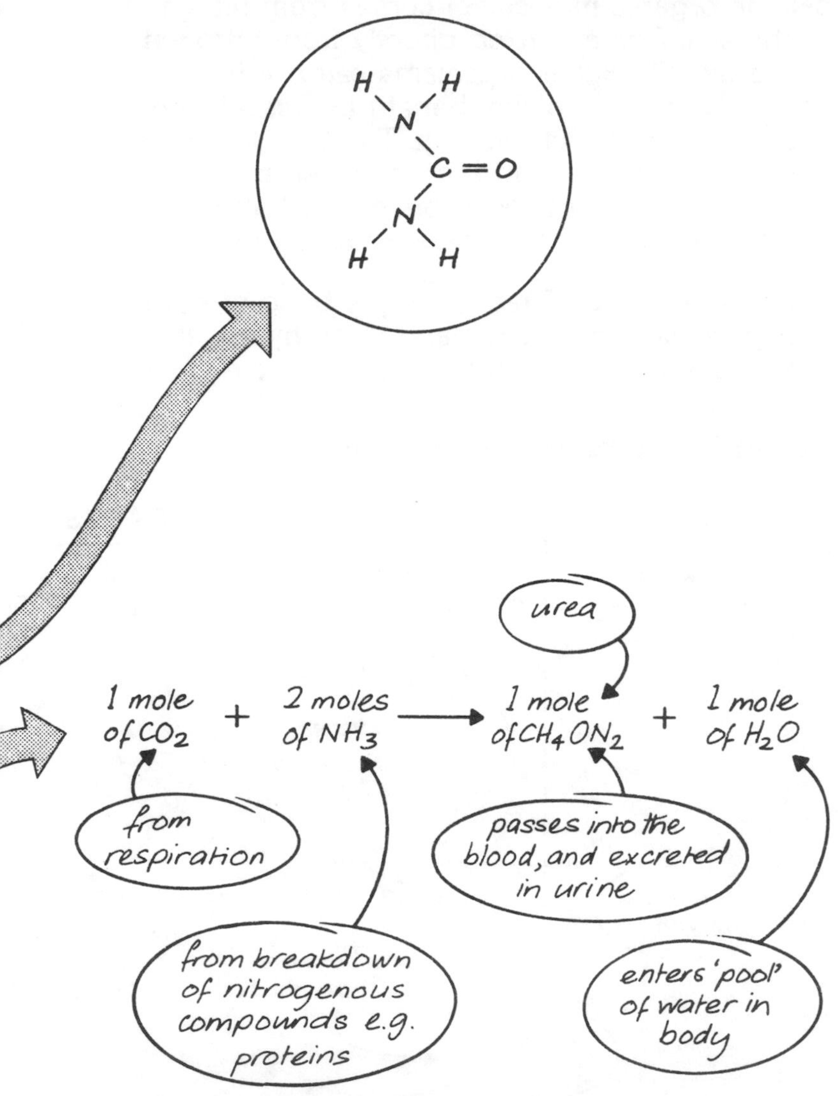
H
H
N
C = O
N
H
H
urea
1 mole of CO2
+
2 moles of NH3
1 mole of CH4ON2
+
1 mole of H2O
from respiration
passes into the blood, and excreted in urine
from breakdown of nitrogenous compounds e.g. proteins
enters 'pool' of water in body

The mechanism is much more complicated than this. **It requires energy**. Like so many metabolic reactions **it is cyclic**, involving amino acids as intermediate carrier molecules. It is called **the urea cycle**.

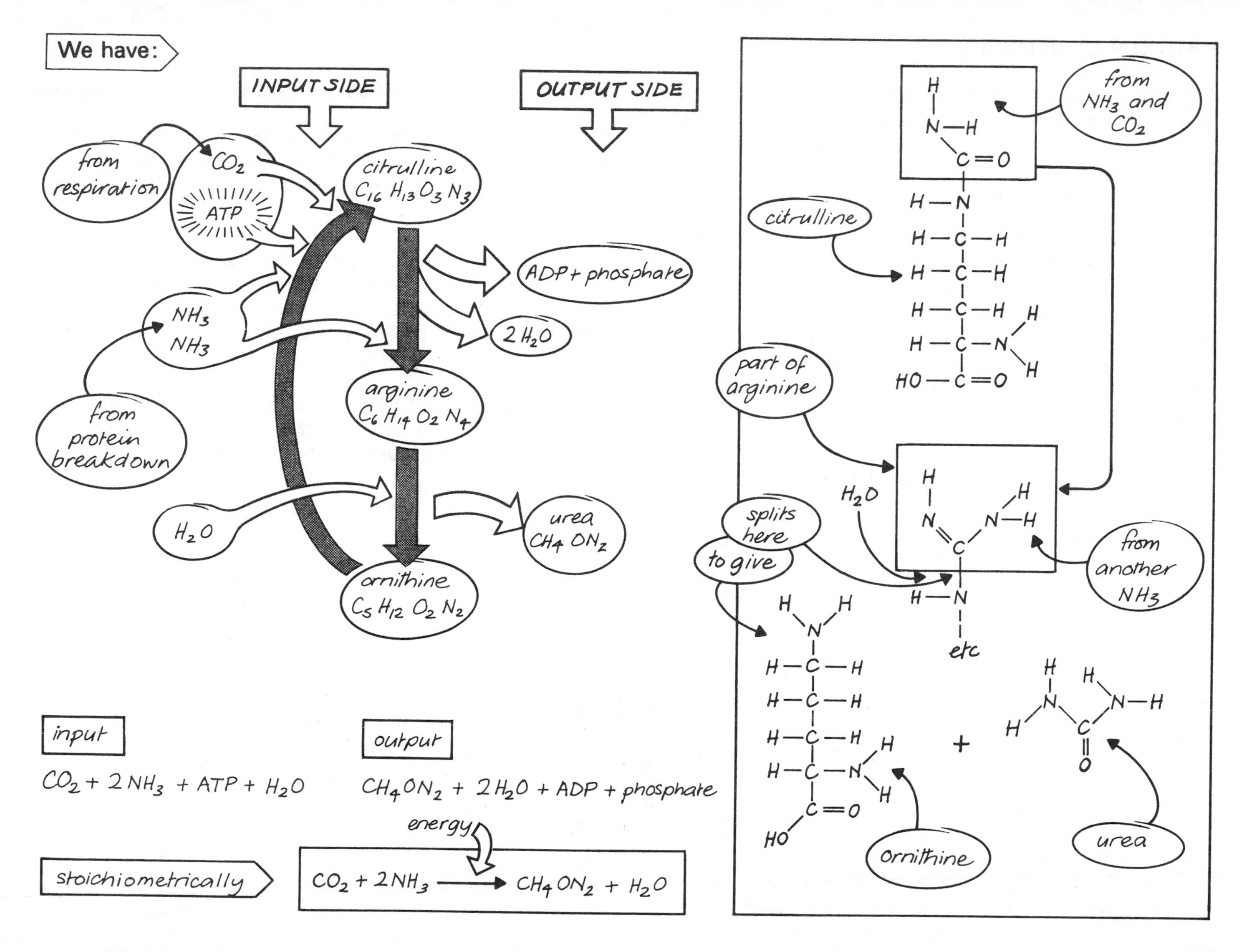
We have:
INPUT SIDE
OUTPUT SIDE
from respiration
CO_2
ATP
citrulline $C_6H_{13}O_3N_3$
ADP + phosphate
NH_3
NH_3
$2H_2O$
from protein breakdown
arginine $C_6H_{14}O_2N_4$
H_2O
urea CH_4ON_2
ornithine $C_5H_{12}O_2N_2$
input
$CO_2 + 2NH_3 + ATP + H_2O$
output
$CH_4ON_2 + 2H_2O + ADP + phosphate$
energy
stoichiometrically
$CO_2 + 2NH_3 \rightarrow CH_4ON_2 + H_2O$
from NH_3 and CO_2
citrulline
part of arginine
H_2O
splits here
to give
from another NH_3
etc
ornithine
urea

PROTEIN SYNTHESIS

If proteins are the 'stuff of life' then elephant protein must be the basic stuff of elephants, human protein of man (as against elephants), and oak tree protein of oaks (as against elephants, man or any other living things, for that matter).

This means that the type of organism is pre-set by the types of proteins of which it is made. We know that the development of each organism is controlled by the nuclei of its cells, so it follows that the pattern of proteins is contained in the cell nucleus, and we now know that **the pattern of each protein in each cell of every organism is contained in its cell nuclei**. The material on which this pattern is imprinted is DNA.

DNA contains the imprint of the organism, and hence the imprint of every protein in that organism.

But DNA is not itself a protein, nor does it resemble a protein at all, chemically. So the pattern, whatever it is, on DNA must be **translated** into protein structure in some way.

We have already seen (page 10) that **once the primary structure of a protein has been settled, the secondary and tertiary structure happens spontaneously by H-bonding and by interaction of the amino acid side groups.**

So the translation of pattern from DNA to protein requires only the order of the amino acid sequence (the primary structure) to be controlled.

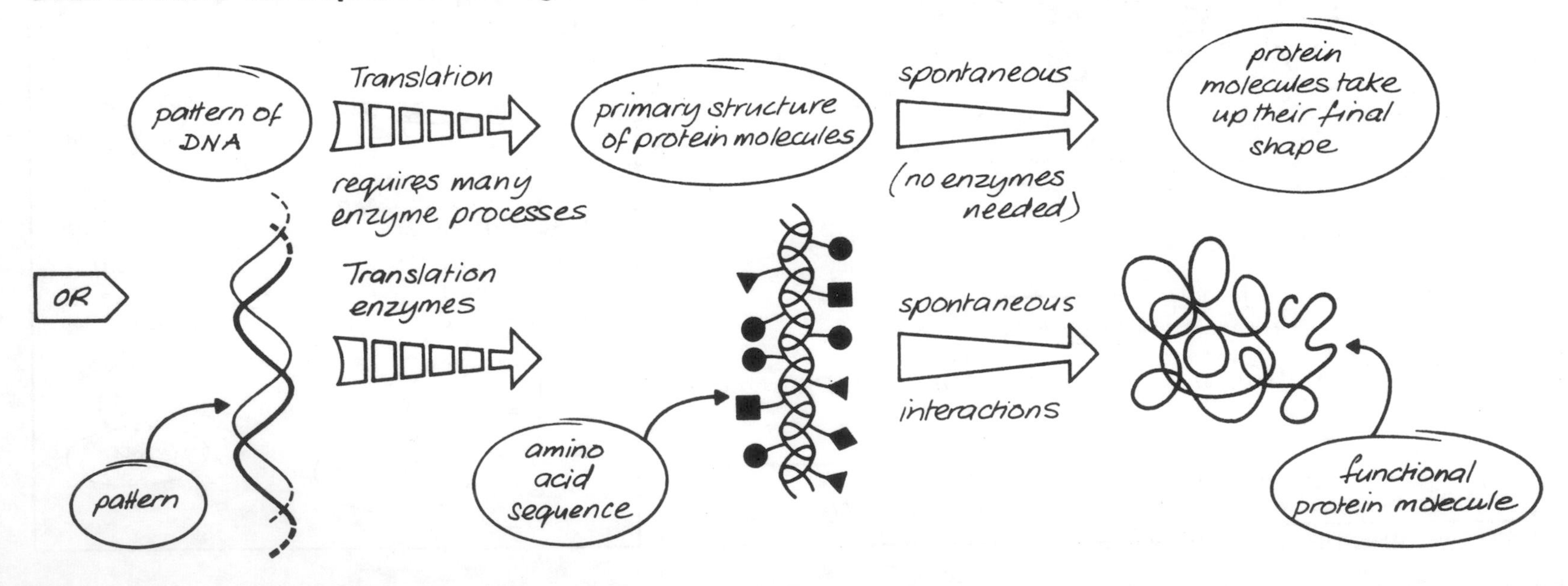
pattern of DNA
Translation
requires many enzyme processes
primary structure of protein molecules
spontaneous
(no enzymes needed)
protein molecules take up their final shape
OR
Translation enzymes
spontaneous
interactions
pattern
amino acid sequence
functional protein molecule

Some basic questions

(1) The pattern of DNA must depend on the order of nucleotides in its molecule – but how can only 4 different types (dAMP, dTMP, dGMP, dCMP) contain this pattern, or **code**, for 20 different amino acids?

– For instance: if each nucleotide in the DNA chain were translated into one amino acid in a protein chain, proteins would be made up of various combinations of only 4 types of amino acid.

We must conclude that the 'information' contained by DNA **is not** a one nucleotide / one amino acid code. What is it?

(2) How does the code, which is on DNA in the nucleus, get to the cytoplasm which is the place where most protein synthesis happens, i.e.

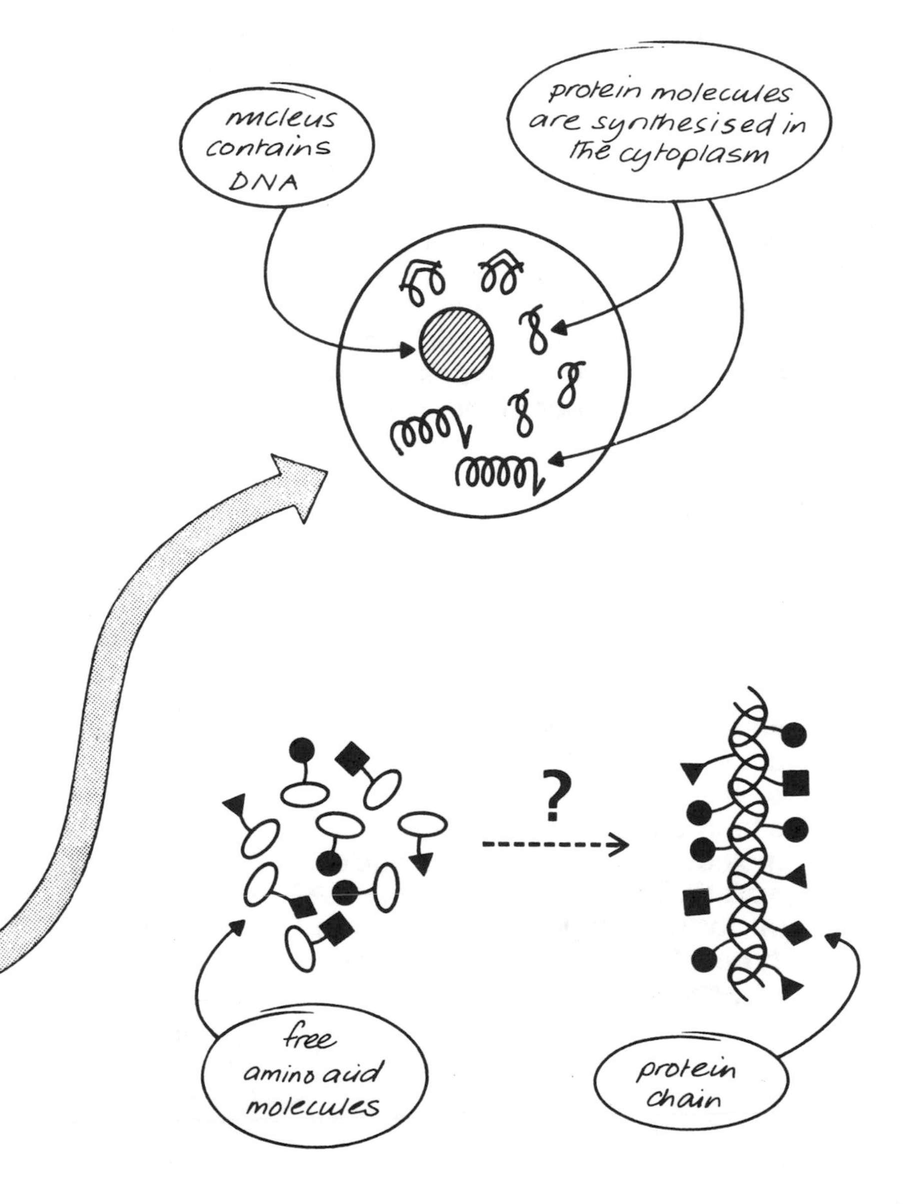
nucleus contains DNA
protein molecules are synthesised in the cytoplasm
?
free amino acid molecules
protein chain

(3) How does protein synthesis actually happen?

The answers

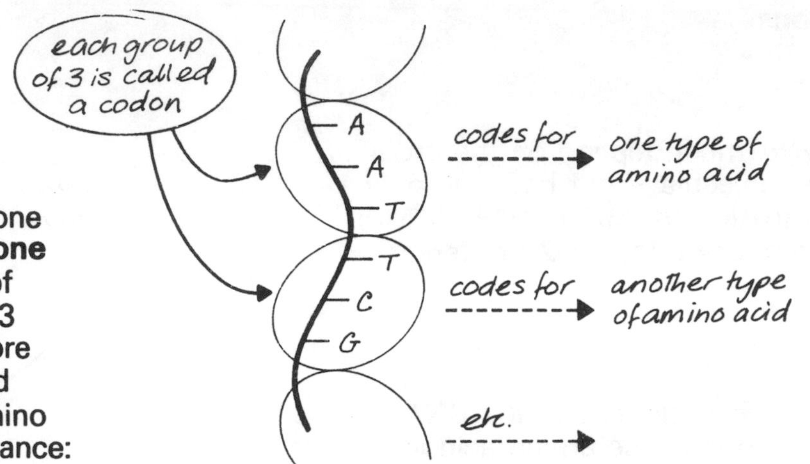

each group of 3 is called a codon
A
A
T
T
C
G
codes for
one type of amino acid
codes for
another type of amino acid
etc.
Note – only one strand of the double helix is used for the code.

(1) The code is not one nucleotide/one amino acid, but **3 nucleotides/one amino acid**. The total number of combinations of 4 things taken 3 at a time is 64. So these are more than enough code 'words' (called **codons**) for the 20 types of amino acids found in proteins. For instance:

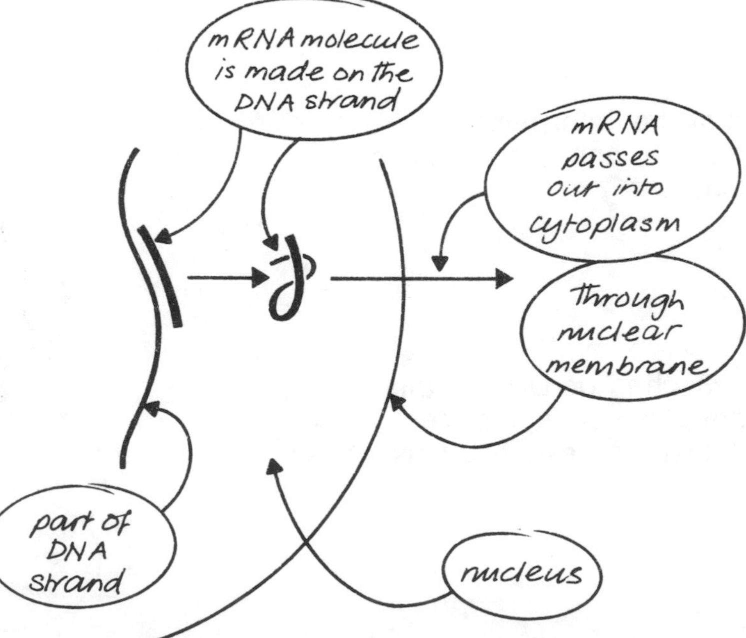

mRNA molecule is made on the DNA strand
mRNA passes out into cytoplasm
through nuclear membrane
part of DNA strand
nucleus
Note – as each mRNA molecule is made on the DNA, it contains a pattern (code) imposed on it by that region of the DNA. Each codon is again a nucleotide 'triplet.'

(2) **Messenger** molecules carry the code from DNA in the nucleus to the protein-synthesising site in the cytoplasm. These molecules are themselves nucleic acid (RNA) and are called **mRNA** (m for messenger). **There is one type of mRNA for each type of protein.**

(3) The answer to this question requires several steps:

(a) each amino acid molecule is activated with ATP.

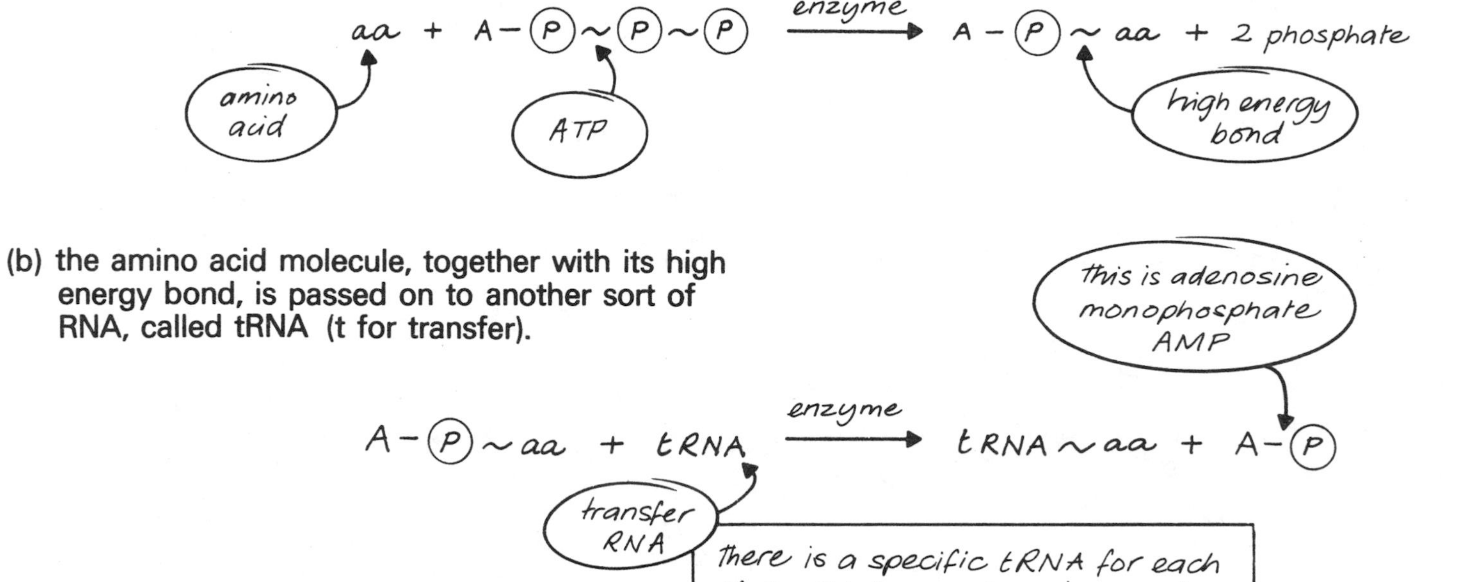
aa + A—(P)~(P)~(P)
enzyme
A—(P)~aa + 2 phosphate
amino acid
ATP
high energy bond
this is adenosine monophosphate AMP
A—(P)~aa + tRNA
enzyme
tRNA~aa + A—(P)
transfer RNA
there is a specific tRNA for each of the 20 types of amino acid.

(b) the amino acid molecule, together with its high energy bond, is passed on to another sort of RNA, called tRNA (t for transfer).

(c) the amino acid is transferred to a growing protein chain.

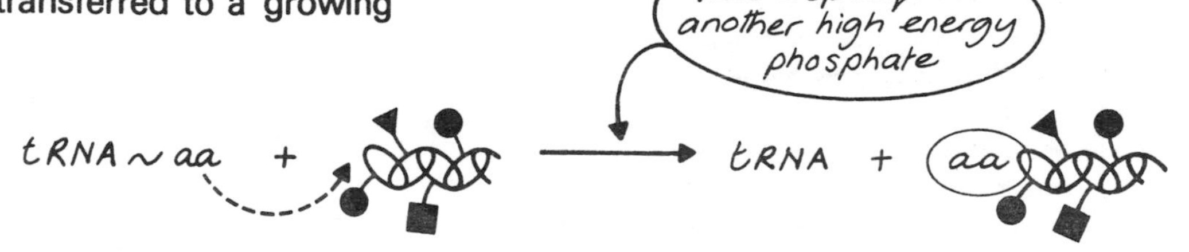
this step requires another high energy phosphate
tRNA~aa +
tRNA +
aa

How does the 'right' amino acid get chosen for this step?

Tiny bodies in the cytoplasm, called **ribosomes**, attach themselves to messenger molecules, **and 'choose' the right tRNA~aa while reading out the (triplet) code**. They synthesise the protein chain while they do so.

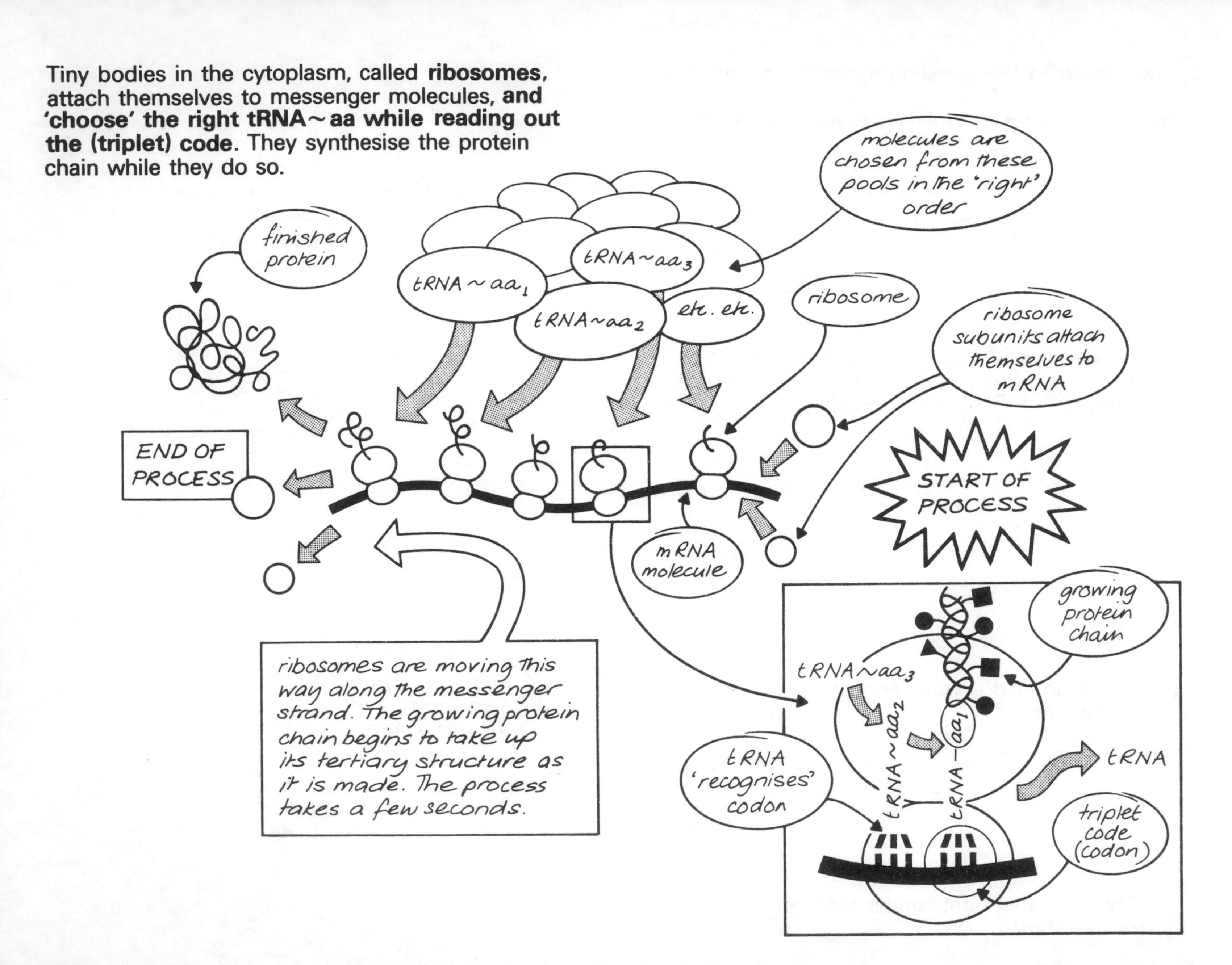

molecules are chosen from these pools in the 'right' order
finished protein
tRNA~aa$_1$
tRNA~aa$_3$
tRNA~aa$_2$
etc. etc.
ribosome
ribosome subunits attach themselves to mRNA
END OF PROCESS
START OF PROCESS
mRNA molecule
ribosomes are moving this way along the messenger strand. The growing protein chain begins to take up its tertiary structure as it is made. The process takes a few seconds.
growing protein chain
tRNA~aa$_3$
tRNA~aa$_2$
tRNA—aa$_1$
tRNA
tRNA 'recognises' codon
triplet code (codon)

Note 1

Ribosomes contain yet another RNA, rRNA (r for ribosomal). About half their weight is protein and half rRNA. They are so small, that we can consider them to be large 'mixed' molecules with a 'molecular weight' of about 3 million.

Note 2

The mRNA is used to make many many protein molecules, each one exactly alike. A different mRNA will make a different type of protein, but each one of those protein molecules will be exactly alike also.

Summary of the types of RNA

symbol	name	function
mRNA	'messenger'	to carry code from DNA
tRNA	'transfer'	to adapt the amino acids to the code
rRNA	'ribosomal'	part of ribosome structure

The synthesis of a protein requires much energy

We have seen that the addition of an amino acid to a protein chain requires 3 high energy phosphate bonds, 2 at the activation stage (to make A–Ⓟ~aa) and one at the peptide-bond stage.

So for a protein with 200 amino acids in its chain:

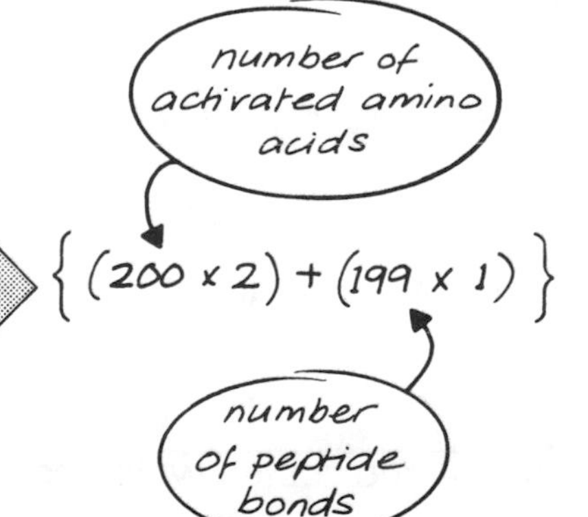
number of activated amino acids
{(200 x 2) + (199 x 1)}
number of peptide bonds

high energy phosphate bonds are required [1 high energy phosphate bond contains 30 kJ free energy/mole]

i.e. 599×30 = about 18,000 kJ per mole of protein (this is actually equal to the amount of energy needed to lift 20 kg of protein (i.e. 1 mole) to the top of St. Pauls!!

The free energy **contained** in the peptide bond is 21 kJ/mole. So 199 bonds contain $199 \times 21 \triangleq 4{,}200$ kJ/mole of protein.

The efficiency of energy use for peptide bond formation from free amino acids is therefore

$$\frac{4{,}200}{18{,}000} = 0.23 = \mathbf{23\%}$$

This appears to be rather poor efficiency, but it does mean that the 'energy pressure' **towards** protein synthesis is very high. This is not surprising as protein synthesis is such an important aspect of cell activity. The high energy phosphate bonds are synthesised (as ATP) during respiration (or photosynthesis by green tissue in the light) and are part of the energy requirement for the life of living organisms. (Actually 2 ATP molecules are used for amino acid activation and 1 GTP molecule is used during peptide bond formation.)

Summary of protein synthesis

– a summary of this complex process is useful.

(1) DNA contains a code for the sequence of amino acids.

(2) DNA can 'make' more DNA for daughter nuclei during cell division.

(3) DNA also 'makes' mRNA molecules which contain the code: this code is based on triplets of nucleotides, each triplet coding for one amino acid.

(4) mRNA molecules pass from the nucleus to the cytoplasm.

(5) In the cytoplasm specific enzymes 'activate' amino acids: tRNA-amino acid complexes are the end results.

(6) Ribosomes present in the cytoplasm attach themselves to the messenger molecules.

(7) Ribosomes pass along each messenger molecule, selecting tRNA-amino acid molecules according to the (triplet) code.

(8) The protein molecule 'grows' from the ribosome and starts to take up its 3 dimensional shape spontaneously as it grows.

(9) After a few seconds the molecule is complete and is discharged into the cytoplasm. The ribosome also 'falls off' the messenger as two subunits, able to restart the process again.

(10) The energy relationships are such that there is a powerful **drive** towards protein synthesis.

Simple organic chemical groups found in biochemistry and used in this book

formula	usually written as	name	some properties
H–C(H)(H)–	CH_3–	methyl	inert, fat soluble
–C(H)(H)–	–CH_2–	methylene	inert, fat soluble
–O–H	–OH	hydroxyl	polar, water soluble
found as:			
–C(H)(H)–OH	–CH_2OH	primary alcohol	↑ more reactive
–C(H)–OH	>CHOH	secondary alcohol	
≥C–OH	≥C–OH	tertiary alcohol	
–S–H	–SH	thiol	reactive, polar, water soluble
–C(H)=O	–CHO	aldehyde	reactive oxidising agent (reduced to –CH_2OH) water soluble

model: C atom, H atom, O atom, S atom; $109\frac{1}{2}°$; 120°; 120°; etc.

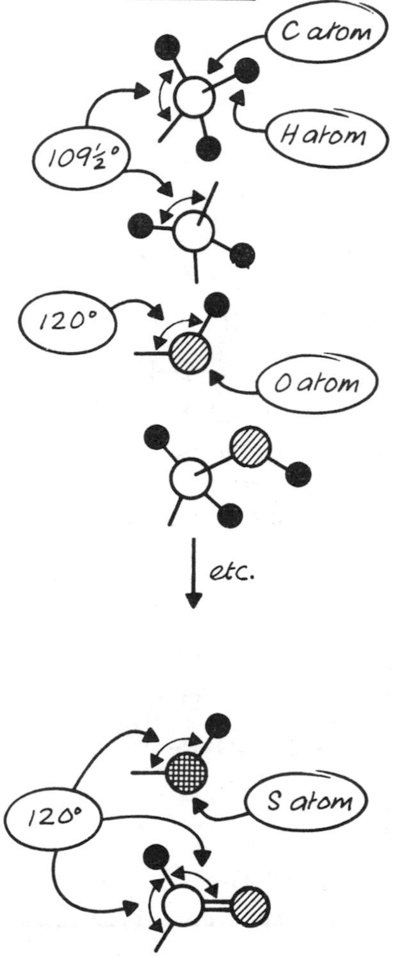

formula	usually written as	name	some properties	model
$>C=O$	$>CO$	carbonyl	reactive (reduced to $>CHOH$) water soluble	
$-C(=O)-O-H$	$-COOH$	carboxyl	reactive acidic (ionised to $-CO_2^{\ominus}$ and H^+) water soluble	
$-N(H)H$	$-NH_2$	amino	reactive basic ($NH_2 + H^+ \longrightarrow NH_3^+$)	N atom; 120°
$-P(=O)(O-H)-O-H$	H_2PO_4-	phosphate	ionised to $PO_4^{2\ominus} + 2H^+$ water soluble	P atom; 109°

usually as organic phosphate ⇩

formula	usually written as	name	some properties	model
$-C-O-P(=O)(O^{\ominus})-O^{\ominus}$	–C–O–(P)	monophosphate		
$-C-O-P(=O)(O^{\ominus})-O-P(=O)(O^{\ominus})-O^{\ominus}$	–C–O–(P)–(P)	diphosphate	↓ more energy	bonds between two phosphates have high energy
$-C-O-P(=O)(O^{\ominus})-O-P(=O)(O^{\ominus})-O-P(=O)(O^{\ominus})-O^{\ominus}$	–C–O–(P)–(P)–(P)	triphosphate		

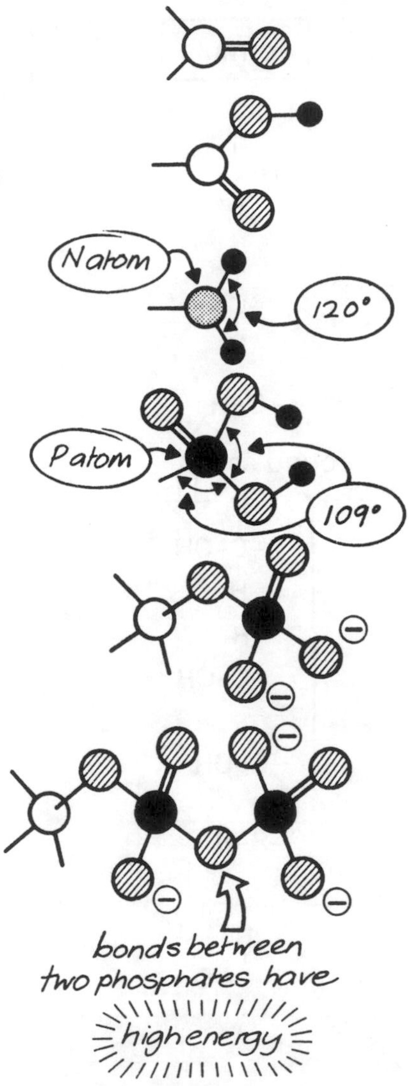

Chemical structures of some of the compounds mentioned in this book

(and not given in the text)

ATP (Adenosine triphosphate)

Used as 'energy currency' in many metabolic processes.

ADP (Adenosine diphosphate)
plus a phosphate ion, plus free energy.

AMP (Adenosine monophosphate)
plus a phosphate ion, plus free energy.

Cytochromes

These proteins participate in hydrogen (electron) transfer reactions; they have prosthetic groups which contain **iron** bound to a class of organic compounds called **porphyrins**.

A series of different cytochrome molecules often form a 'chain' along which electrons pass and change their energy state (e.g. respiration, photosynthesis).

Chlorophylls have similar prosthetic groups to cytochromes, but the **iron** molecule is replaced by **magnesium**.

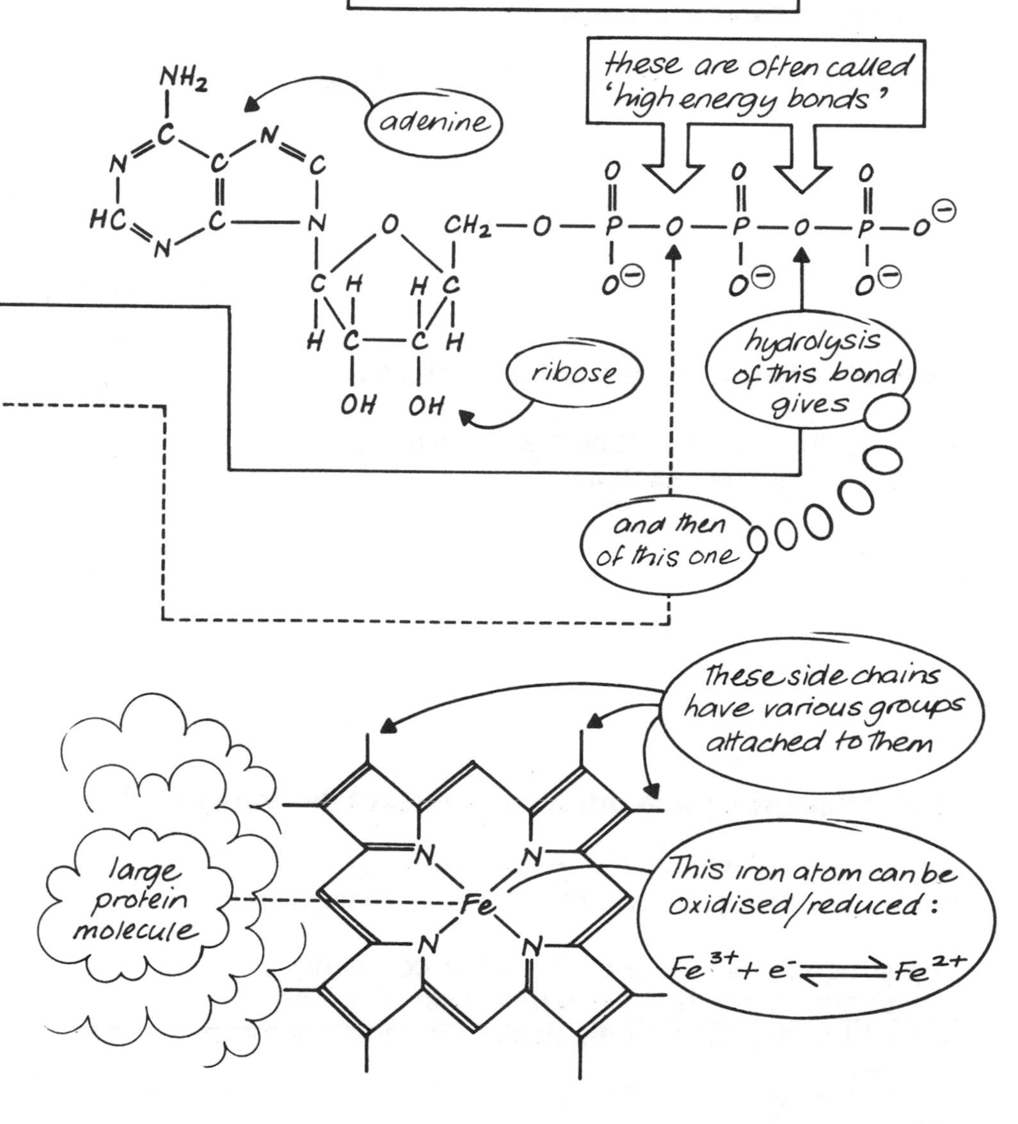

adenine
these are often called 'high energy bonds'
ribose
hydrolysis of this bond gives
and then of this one
these side chains have various groups attached to them
large protein molecule
This iron atom can be oxidised/reduced:
$Fe^{3+} + e^- \rightleftharpoons Fe^{2+}$

FAD (Flavine adenine dinucleotide)

A coenzyme which participates in hydrogen (electron) transfer reactions.

Flavoprotein is a protein with FAD attached.

Flavoprotein is a protein with FAD attached

This reduction is easily reversed.

NAD (Nicotinamide adenine dinucleotide)

A coenzyme which participates in hydrogen (electron) transfer reactions.

Note: *In this book, for reasons of clarity, we have used the symbols NAD and $NADH_2$. It is more correct to use NAD^+ and NADH.*

This reduction is easily reversed.

NADP (Nicotinamide adenine dinucleotide phosphate)

is a similar coenzyme. It contains an extra phosphate group on one of the ribose molecules.

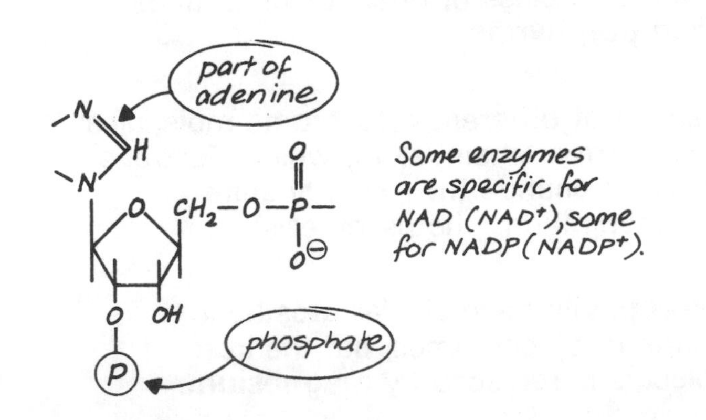
part of adenine
phosphate
Some enzymes are specific for NAD (NAD+), some for NADP (NADP+).

The structures of most of the other compounds mentioned in the text are either given there or should be known from elementary organic chemistry.

Index